A Garden for the President

A History of the White House Grounds

A Garden for the President

A History of the White House Grounds

BY JONATHAN PLISKA
PRINCIPAL PHOTOGRAPHY BY BRUCE M. WHITE

THE WHITE HOUSE *HISTORICAL ASSOCIATION*

WHITE HOUSE HISTORICAL ASSOCIATION is a nonprofit educational association founded in 1961 for the purpose of enhancing the understanding, appreciation, and enjoyment of the Executive Mansion. All proceeds from the sale of the association's books and products are used to fund the acquisition of historic furnishings and art work for the permanent White House collection, assist in the preservation of public rooms, and further its educational mission.

This book has been brought to publication through the generous assistance of the Hon. Walter H. Annenberg Publications Fund

Vice President of Publishing and Executive Editor: Marcia Mallet Anderson
Editorial Director: Fiona Griffin
Production Manager: Lauren A. Zook
Editorial Specialist: Olivia Sledzik
Editorial Assistant: Kristin Skinner
Consulting Editor: Ann Hofstra Grogg
Designer: Pentagram

First Edition
10 9 8 7 6 5 4 3 2 1
Library of Congress Control Number: 2015957305
ISBN 978-1-931917-55-1
Printed in Italy

Dedicated by the White House Historical Association

to the National Park Service on the

Occasion of the Centennial of its Creation

CONTENTS

FOREWORD
XI

PREFACE
XIII

ACKNOWLEDGEMENTS
XIV

INTRODUCTION
1

THE GROUNDS
24

The North Grounds
26
Front Lawn / North Entrance
Jefferson Statue / North Fountain

The South Grounds
44
The President's Backyard / Mystery of the Mounds / Reshaping the Grounds / Winding Roadways / Fences / Fountains / South Vista / Children's Garden

The West Gardens
76
First Ladies' West Gardens / Mid-Century West Garden / Kennedy Rose Garden / An Outdoor Oval Office

The East Gardens
98
First Ladies' East Gardens / Mid-Century East Garden / Jacqueline Kennedy Garden

Visiting the White House
110
East Entrance

THE USES OF THE PRESIDENT'S GROUNDS
116

Public Access
118
President's Grounds as the Public Grounds / Weekly U.S. Marine Band Concerts / Annual White House Easter Egg Roll

Private Uses by First Families
144
Parties on the South Grounds / Presidential Recreation / First Children and Pets at Play

White House Grounds During Wartime
172
War of 1812 / Civil War / World War I / World War II

WHAT GROWS ON THE PRESIDENT'S GROUNDS
190

The Trees as Living Witnesses
192
History of Tree Planting at the White House / Oldest Trees / Commemorative Trees / John Quincy Adams's Historic Trees / Russo-American Oak / Franklin D. Roosevelt and Thomas Jefferson / Jackson Magnolias

Food for the President's Table
244
White House Kitchen Garden / Demonstration Gardens

Ornamental Gardening
256
First White House Flower Garden / Andrew Jackson Downing's Fence / Gardening Under Glass / Great Glasshouse Complex

EPILOGUE
284

CREATORS
288

SOURCES, RESOURCES AND NOTES
296

ILLUSTRATION CREDITS
303

INDEX
304

FOREWORD

In my more than forty years of service at the White House, I have witnessed first-hand the reality of First Lady Abigail Adams's astute observation that the President's House is situated "in a beautiful spot, capable of every improvement." She expressed this sentiment in a letter to her daughter in November 1800, shortly after moving into the yet unfinished mansion, still surrounded by mud and construction debris. "We have not the least fence, yard, or other convenience, without," she wrote, yet she concluded the letter with the happy thought, "the more I view it, the more I am delighted with it." After two hundred years of continued devotion and labor, and through the undertaking of ambitious plans directed by such horticultural giants as Frederick Law Olmsted Jr., I think Mrs. Adams would be pleased to see that, today, the beautifully landscaped grounds encompass just as she predicted, every improvement—fences and gates, drives and vistas, fountains and ponds, and flower gardens and groves of trees, to name just a few. And over time, but now consigned to memory, there have also been greenhouses and stables, and colonial, Beaux-Arts, and even sculpture gardens. Change is as constant in the garden as are the daily routines that have been carried on for two centuries, planting with the seasons, caring for the trees, and ensuring that the green grass that serves as a presidential stage is always green.

Over the years, I have witnessed the optimism of Abigail Adams in every commemorative tree planting and I have shared in the sadness when storms or age bring down venerable old trees—trees that touched the past—beginning their life perhaps in time to witness troop encampments during the Civil War, suffragettes at the gate, or the first automobiles on the drive.

The many visitors I accompany through the gardens want to know as much—or more—about the human history of the grounds as the horticultural history. I am often asked, "Who was married in the Rose Garden?" "Where was Amy Carter's tree house?" and "Where are the State Arrival ceremonies held?" Most such questions have definite answers, others less so. The answers to "Who planted the Jackson magnolia?" and "When were the Jefferson mounds created?" are not as straightforward as you might guess. As Jonathan Pliska's text reveals, there is so much we have learned about the President's Grounds from documents of the past, and still some mysteries to solve. By the way, there is something I could add to Jonathan's chapter on the Mounds: They are the best place to roll Easter Eggs and the best place to view the fireworks on the Fourth of July.

You can't work in the White House gardens for as long as I have without acquiring favorite places and seasons, a repertoire of stories to tell, and some insider secrets. I am especially fond of the annual spring and fall garden tours, when the public is welcomed to tour the South Grounds. The tradition was begun by First Lady Pat Nixon in 1972, my first year of working at the White House. Having now spent my career maintaining the gardens, it is especially rewarding to see the crowds enjoy the tours and to show them such special features as 8,000 Oxford tulips surrounding the South Fountain, which, in combination with a border of grape hyacinth, create an appropriately red, white, and blue element within the South Vista. Every president has walked the grounds, and every one has left something behind—perhaps a vision, a plan, a tree, a fence, or a swimming pool. I have observed that even first time visitors can sense this rich history and I see they take great pleasure in the opportunity to experience the grounds for themselves.

I have often wished for a book that would tell the whole story of life and growth in the garden from the beginning until today, a book that would supplement the experience of a spring garden tour or provide a virtual tour to those who may not be able to visit in person. In the following chapters Jonathan Pliska has accomplished this and Bruce White, with his camera, has captured the spirit of the landscape from a gardener's perspective. They love the subject, and the reader will love it too.

Dale Haney
Grounds Superintendent, The White House

PREFACE

Since its founding more than half a century ago, the White House Historical Association has endeavored to enhance the understanding, appreciation, and enjoyment of the Executive Mansion, but the focus of our educational programs and publications has never been limited to marble halls and State Rooms. The historic grounds that surround the President's House have served as a stage for American History for more than two hundred years, and the study of these grounds is naturally integral to our programming.

Although the early presidents knew their vision for the eighty-five acres of President's Park would not be fully realized in their lifetimes, they approached landscaping and tree planting with selfless optimism for the future. I am sure that President Thomas Jefferson, who drew ambitious plans, President John Quincy Adams, who gathered and planted acorns, and President Andrew Jackson, who installed the curving drives, would be pleased to see the mature groves of trees and formal landscape at the heart of the nation's vibrant Capitol City today. Every president and first family who lived in the house contributed in some way to the evolution and use of the grounds. We can still see the red Japanese maple planted by Frances Folsom Cleveland, the Children's Garden conceived by First Lady Lady Bird Johnson, the swimming pool built by President Gerald Ford, and the Kitchen Garden installed by First Lady Michelle Obama, to name just a few examples of this visible legacy. With the help of Jonathan Pliska's thorough scholarship, so beautifully illustrated in the following chapters of this book, we can also imagine President Zachary Taylor mingling with the crowd on the South Lawn during a performance of the U.S. Marine Band; the Union troops encamped on the grounds to protect President Abraham Lincoln during the Civil War; two hundred happy children, welcomed by President Rutherford B. Hayes, rolling their eggs on the White House grounds; late nineteenth-century conservatories filled with orchids; high society enjoying First Lady Edith Roosevelt's gracious garden parties; a dairy cow named Pauline Wayne grazing on President William Howard Taft's front lawn; President John F. Kennedy's Rose Garden under construction; and President and Mrs. George H. W. Bush overseeing Millie and her puppies frolicking on the South Lawn.

The oldest continually maintained landscape in the nation, the White House Grounds are the responsibility of the National Park Service, which celebrated its centennial this year. This book is the Association's recognition of this milestone of stewardship and a celebration of the ongoing commitment of the National Park Service to fulfilling its mission of ensuring the "continued protection, preservation, serviceability, and use" of the National Parks across the nation.

Stewart D. McLaurin
President, White House Historical Association

ACKNOWLEDGMENTS

For Mom, Dad, Luke, Bill, and Barry—you are all loved, and you are all missed.

This book is being published now, in the year 2016, to celebrate the one hundredth anniversary of the founding of the United States National Park Service. In 1933, the NPS took over administration of all federal parkland in Washington, D.C., including the White House Grounds. Since the 1960s, the NPS has also been responsible for the care, upkeep, improvement, and preservation of the landscape. Today, a team of NPS gardeners and maintenance professionals are employed full time to tend the President's Grounds. Others are called in as needed to assist with special projects and still more work off-site to ensure that everything runs smoothly. All have my sincere respect and deepest gratitude for their work.

I began studying the White House Grounds in 2008, as a contractor for the NPS. Paul Dolinsky, head of the Park Service's Historic American Landscapes Survey program, secured funding from the White House Historical Association for me to study the history of the grounds in a full and comprehensive manner. This opportunity allowed me to venture down previously unexplored avenues of research that led to new and exciting discoveries. NPS architect Jason McNatt graciously lent his time and talent on many occasions, enhancing my ability to trace the long, complicated development of the President's Grounds over its history. Finally, the staff of the NPS Liaison to the White House shared their expertise and resources, without which the project might never have gotten off the ground.

The late Dr. William Bushong, chief historian for the White House Historical Association, readily offered his encouragement and advice whenever needed, but unfortunately passed away during the editorial stages of this book. Please let this publication serve as a posthumous reminder of his kindness and professionalism.

My great thanks are owed to Stewart McLaurin, president of the White House Historical Association, who decided the time was right for a new scholarly work on the subject, and enthusiastically gave his support. Dr. William Seale, author of authoritative books and articles on the White House and its grounds, reviewed my work, provided welcome feedback, answered questions, and, perhaps most importantly of all, kept me from losing perspective. In the White House Historical Association's publications department, Marcia Anderson, the vice president for publications and executive editor, led the way during the book's development. Lauren Zook, Fiona Griffin, Olivia Sledzik, and Kristin Skinner organized myriad editorial and production details. Ann Hofstra Grogg, consulting editor, shaped my manuscript into a book, making the final version so much better than the first, while keeping my vision intact.

I am indebted to the Office of the Curator, the White House—especially to William Allman and Lydia Tederick, for commenting and advising on the final draft and generously offering their vast institutional knowledge. Likewise, Dale Haney, grounds superintendent at the White House, enthusiastically facilitated photography over many years and contributed invaluable suggestions on all things botanical.

I would also like to recognize the creative talent of Bruce White, whose new and original photographs practically jump off the page and pull you into the green spaces of the President's Grounds with a never-before seen perspective. Similarly, Luke Hayman and Ellen Peterson of Pentagram created an absolutely beautiful design for this book.

Finally, thank you to my family and friends for your constant love and support. Most of all to Ellen, my dear wife, my rock, and my source of inspiration. I love you more than anything, and I always will.

Jonathan Pliska

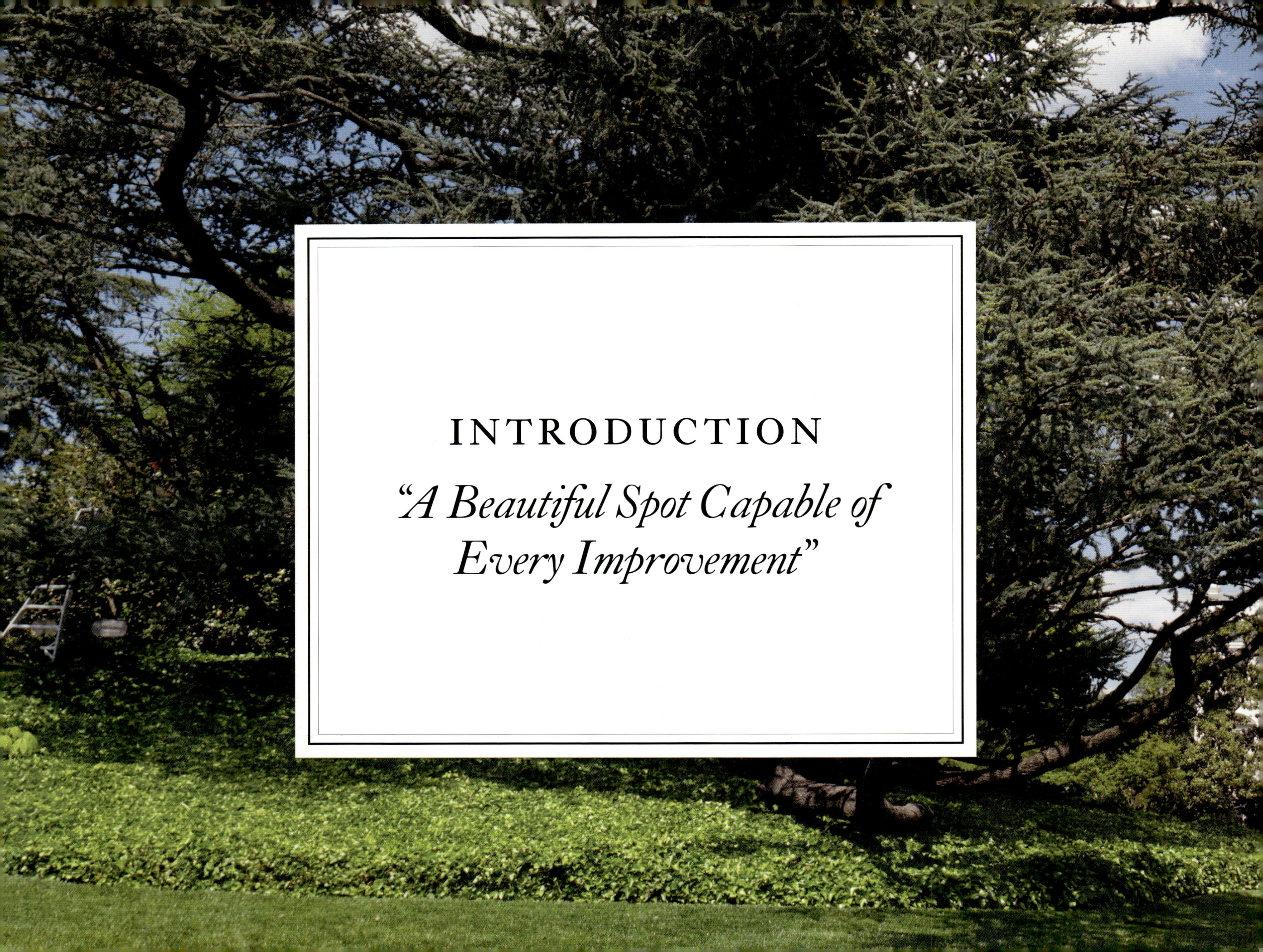

INTRODUCTION

"A Beautiful Spot Capable of Every Improvement"

THE HISTORY OF THE WHITE HOUSE GROUNDS begins nearly two centuries before the construction of the house itself. Sailing up the Potomac River in 1608, Captain John Smith and the members of his exploring party became the first non–Native Americans to lay eyes on the future site of the presidential mansion. Algonquin and Nacotchtankes people already called the area home, and they engaged in agriculture, hunting, and fishing. Subsequent clashes with European colonists forced them off the land, which, by the time of the American Revolution, had been carved up into parcels tended by five generations of English planters. The area was rural, sparsely populated, and quite ordinary, but all that would change.

On July 16, 1790, Congress passed the Residence Act into law. It granted President George Washington authority to select a "district of territory, not exceeding ten miles square, along the Potomac River," as the permanent seat of government of the United States. Six months later the president fixed the boundaries of this federal district, and in September 1791 it was officially named "Columbia," after Christopher Columbus. President Washington received the singular honor of having the capital city bear his name, and so it became the City of Washington within the District of Columbia. Maryland and Virginia ceded the necessary land to the federal government, but at the time this national capital was just a patchwork of forests, fields, orchards, and swamps between the established ports of Georgetown and Alexandria. Congress therefore chose to make Philadelphia the temporary capital and seat of government until December 1, 1800, when the federal government would permanently relocate to Washington, D.C. This decision allowed less than a decade to create a city: to develop a city plan, conduct surveys, produce maps, hire architects and builders, and design and construct the government buildings.

Congress entrusted the details to President Washington, and he in turn selected French-born Pierre Charles L'Enfant to draw the plan for the new federal metropolis. L'Enfant was a talented artist and military engineer who had served under Washington's command during the Revolution. After the war, he stayed on to organize banquets, parades, and pageants to promote the ratification of the Constitution. L'Enfant threw himself into his new assignment. After arriving in Georgetown on a cold spring night in March 1791, he selected sites for government buildings and in his mind formulated a vast web of interconnecting streets. A mere six months later he was pleased to present his city plan to President Washington, who approved it in nearly every respect.

L'Enfant's plan included a Presidential Palace atop a lofty ridge overlooking the Potomac River and set within a large plot of land ranging over approximately 80 acres. Washington himself made the final decision on its placement, selecting a spot just west of L'Enfant's proposed location so that the palace would sit as high as possible. From this point, the view stretched

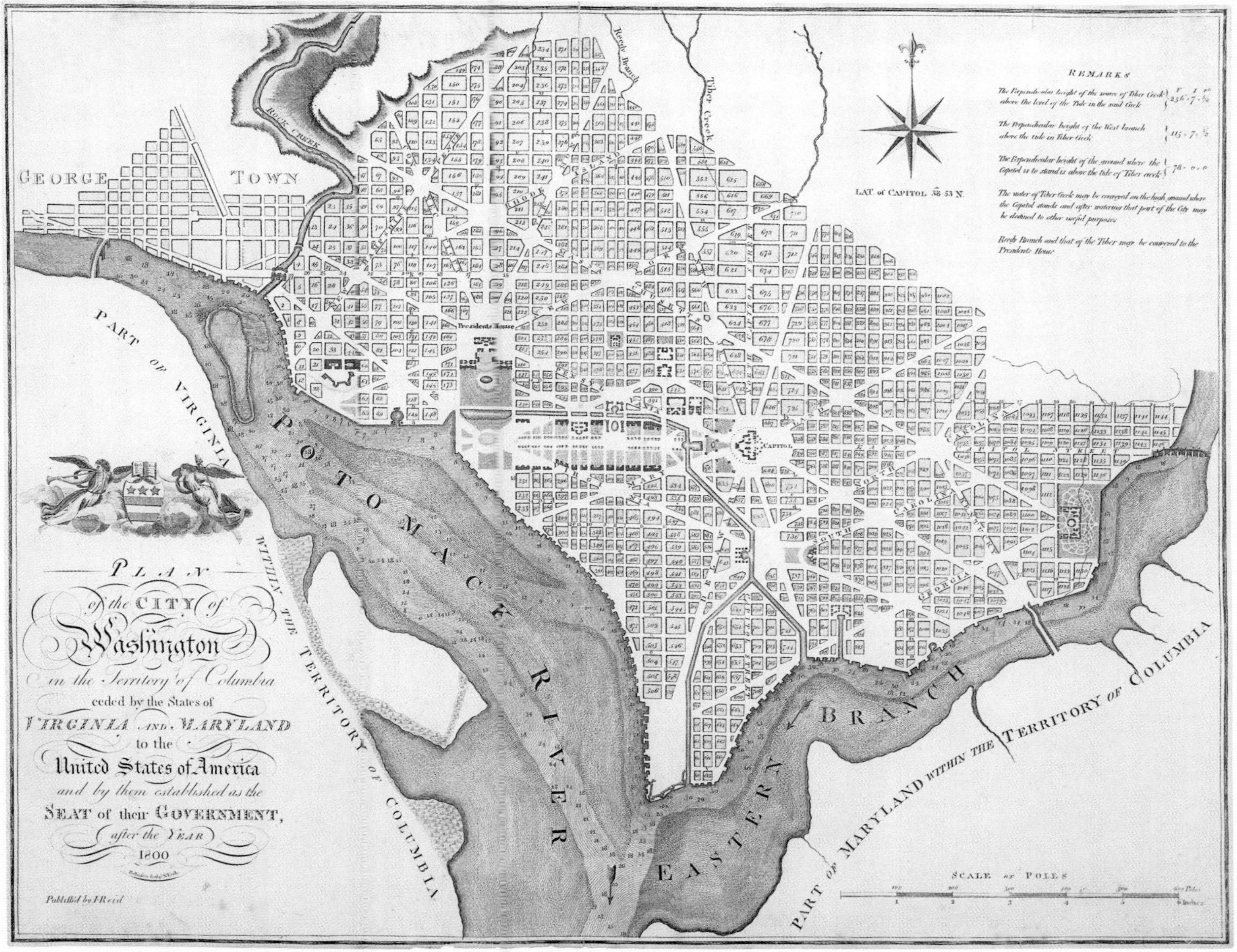

Anchored by the sites of the future U.S. Capitol Building and the President's House, L'Enfant's Plan of the City of Washington *is essentially a grid of intersecting north-south and east-west streets overlaid with diagonal grand avenues that incorporate traffic circles, squares, and other parcels of land reserved for the development of urban parks, memorial spaces, and government offices. Pennsylvania Avenue, one of the largest of these thoroughfares, was intended to link the Capitol and the immense building that L'Enfant christened the "President's Palace."*

Pierre Charles L'Enfant's natural reputation as a "gifted designer of monumental architecture" was established in large part by his transformation of New York's Old City Hall into "Federal Hall" in 1788–89. The impressive neoclassical edifice emblazoned with gilded stars, eagles, and Roman motifs served as the first meeting place for Congress and was considered the finest building in the country. The success of this project contributed to his securing the position as architect of the City of Washington. His profile is seen here on this bas-relief by an unknown artist.

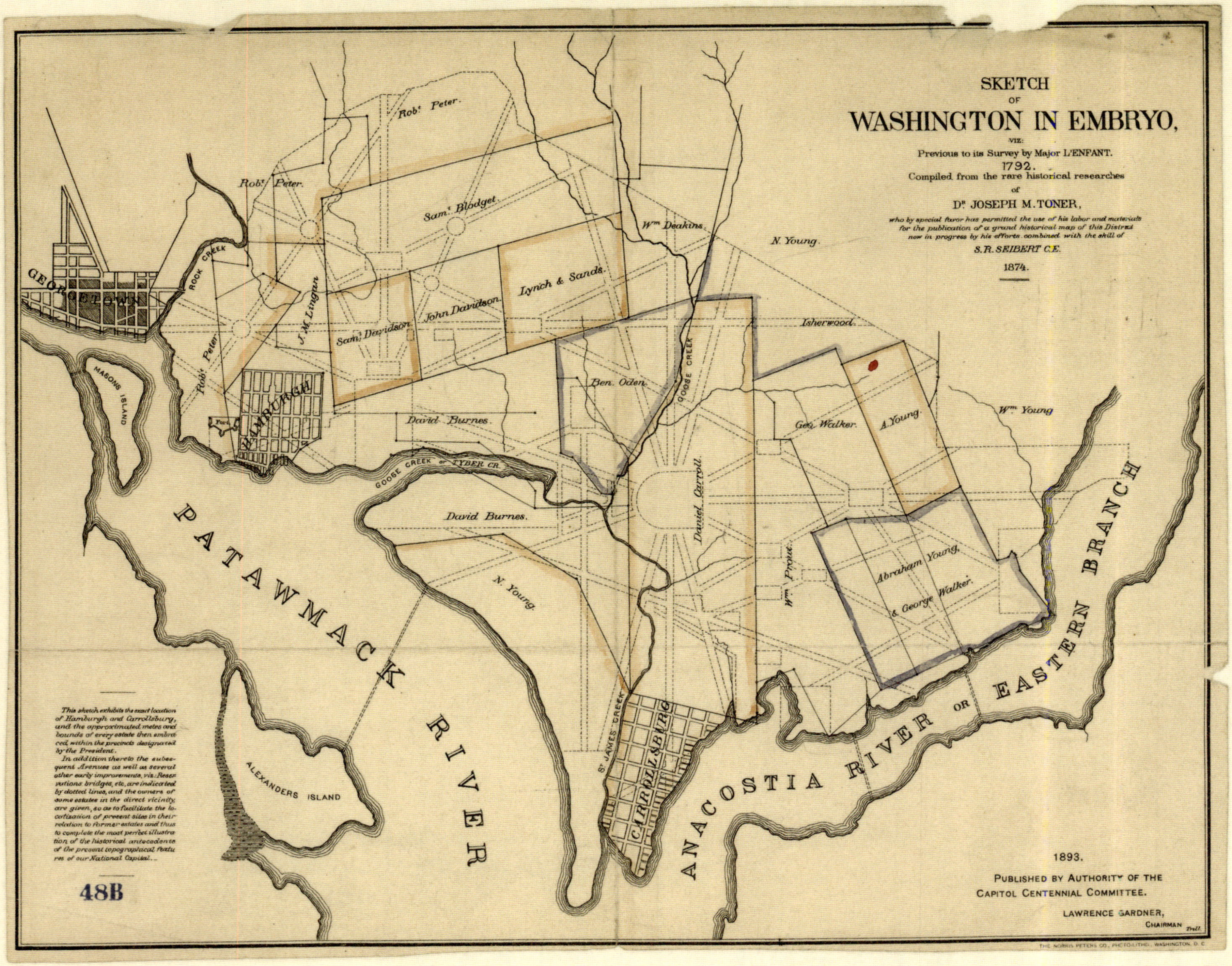
SKETCH
OF
WASHINGTON IN EMBRYO,
VIZ:
Previous to its Survey by Major L'ENFANT.
1792.
Compiled from the rare historical researches
of
Dr. JOSEPH M. TONER,
who by special favor has permitted the use of his labor and materials for the publication of a grand historical map of this District now in progress by his efforts combined with the skill of
S. R. SEIBERT C.E.
1874.
This sketch exhibits the exact location of Hamburgh and Carrollsburg, and the approximated metes and bounds of every estate then embraced within the precincts designated by the President.
In addition thereto the subsequent Avenues as well as several other early improvements, viz: Reservations, bridges, etc, are indicated by dotted lines, and the owners of some estates in the direct vicinity are given, so as to facilitate the localization of present sites in their relation to former estates and thus to complete the most perfect illustration of the historical antecedents of the present topographical features of our National Capital.
48B
GEORGETOWN
ROCK CREEK
MASONS ISLAND
HAMBURGH
Robt. Peter.
Robt. Peter
J. M. Lingan
Saml. Blodget.
Saml. Davidson
John Davidson
Lynch & Sands.
Wm. Deakins.
N. Young.
Isherwood.
Ben. Oden
GOOSE CREEK
David Burnes.
GOOSE CREEK or TYBER CR.
Geo. Walker.
A. Young.
Wm. Young
Daniel Carroll
Wm. Prout.
Abraham Young
& George Walker.
N. Young
PATAWMACK RIVER
ALEXANDERS ISLAND
St. JAMES CREEK
CARROLLSBURG
ANACOSTIA RIVER OR EASTERN BRANCH
1893.
PUBLISHED BY AUTHORITY OF THE
CAPITOL CENTENNIAL COMMITTEE.
LAWRENCE GARDNER,
CHAIRMAN

This 1792 plat, entitled Sketch of Washington in Embryo, *illustrates land ownership prior to the development of the Federal City. Edward Pierce operated an apple orchard and tobacco plantation on a tract that included the current North Grounds of the White House. He quickly sold out to Samuel Davidson, a Georgetown merchant-turned-land speculator, who then resold the parcel to the government at a profit. The area south of the ridge, now the South Grounds of the White House, was owned by a second tobacco planter, the obstinate David Burnes.*

out some 10–12 miles, attracting visitors to the site before construction crews even broke ground.

But first the federal government had to purchase the land, and President Washington appointed three city commissioners to negotiate with the owners, enter into contracts on his behalf, and organize the subsequent building efforts. One of the owners, David Burnes recognized that this was a seller's market and held out for more money. The commissioners countered with a low offer, pointing out that the soil had already been depleted by years of tobacco cultivation and that the swamps were unsuited to any type of farming. Still Burnes stood firm, even after Washington personally involved himself in the negotiations. In the end, the government paid his price.

With the land acquisition completed, L'Enfant laid out the footprint of the President's Palace in December 1791 and supervised a partial excavation of its cellar carried out by African American slaves. It was his last major contribution to the project. Although skilled and industrious, L'Enfant was arrogant, obstinate, and abrasive. After months of fighting with the city commissioners, who were legally L'Enfant's supervisors, President Washington made the difficult decision to dismiss his old friend for the sake of progress.

L'Enfant's plan for the President's Palace disappeared with him. The structure was too large and its surrounding grounds too vast to be made ready by the December 1, 1800, occupancy deadline. Equally important, the regal name was a poor choice for the home of the chief executive of a republic. So the President's Palace became the President's House, and the commissioners announced a public design competition to select its architect. They required that the landscape inform the design: "The site of the building, if the artist will attend to it," read the official announcement, "will of course influence the aspect and outline of his plan and its destination will point out to him the number, size and distribution of the apartments." They also wanted to scale back L'Enfant's plans for the grounds. Commissioner David Stuart warned President Washington that the proposed 82-acre park was out of character with the scaled-back vision for the building: "The ground about the president's house [is] much too extensive. It may suit the genius

The city commissioners, with President Washington's blessing, chose a sketch submitted by James Hoban for the design of the President's House on July 17, 1792. Irish-born architect James Hoban, whose profile in wax is seen here, was likely influenced by Leinster House in Dublin, Ireland. Hoban's revised scheme of 1793 for the North Elevation is seen far left.

As illustrated in c. 2009 by Peter Waddell, in A Vision Takes Form: The White House Under Construction, 1796, *the exterior shell of the President's House is faced with Aquia Creek sandstone quarried in Stafford County, Virginia, while its inner skeleton is comprised of bricks made from soft, red clay soil dug from its surrounding grounds. The brickmaking, which eventually involved three large kilns and a number of smaller, secondary ones, began in October 1793.*

of a despotic government to cultivate an immense and gloomy wilderness in the midst of a thriving city. . . . I cannot think it suitable in our situation." If a king's palace was not appropriate for the president of the United States, then a king-size landscape was equally inappropriate. The winning design for the future White House was submitted by Irish-born architect James Hoban.

Hoban visited the site while drafting his submission, and the president and commissioners evidently felt that the structure he proposed was well suited to the landscape. It was also only about a quarter the size of L'Enfant's palace. Even after Washington added two bays to the design, it was clear Hoban's design would fill only a portion of the excavated ground. So where, exactly, would it be built? President Washington had the final say, and he arrived in early August 1792 to personally stake out the mansion's perimeter. The issue greatly vexed him as he paced back and forth, but eventually he decided to center the North Front of the White House along the northern edge of the excavation, thus ensuring that the confluence of streets to the north would remained intact and precluding the need to alter the city plan. Yet because the house was so much smaller than the palace, this placement moved the President's House out of alignment with the U.S. Capitol along Pennsylvania Avenue. L'Enfant had intended to link the executive and legislative branches of government along this sightline, but the connection was lost before construction of either building began.

The cornerstone of the President's House was laid on October 13, 1792. President Washington, contrary to some accounts, remained in Philadelphia and did not attend the ceremonies. Over the next eight years, the area around the mansion was a busy construction site. Ever-increasing pressure to make the house habitable by December 1, 1800, allowed no time to improve the grounds. Even finalizing the boundaries of the presidential landscape was unresolved in the great rush to complete the house.

Most of the builders lived on-site in small cottages and huts, apparently grouped on what is now the North Lawn. Storehouses, workshops, and a cookhouse appeared as well. The largest of these temporary buildings was the "carpenter's hall," which also doubled as the local Masonic lodge and

In this poster made in c. 1930 for the Pennsylvania Railroad (far left), artist N. C. Wyeth imagined the scene, when on March 14, 1797, former President George Washington made his final inspection of the President's House, still under construction. A crowd gathered to catch a glimpse of him, and the local militia fired a sixteen-gun salute in his honor. It was the only documented reception that Washington ever attended on the grounds, and he never returned to the site.

When President John Adams moved into the President's House, the interior rooms were not yet fully finished, the plastering remained a work in progress, and countless details still required attention. The grounds remained unimproved, save for a small privy. Artist Tom Freeman imagined the desolate scene in this 2000 watercolor (left).

Architect of the Capitol Dr. William Thornton, at the request of the secretary of the navy, made this sketch in 1800 to show how the grounds might be improved with Lombardy poplars. The small face at the bottom resembles Thornton and may be a self-portrait.

served as a church on Sundays. The scene resembled a village, populated by all manner of workers, from skilled tradesmen to slaves, and filled with stacks of building materials. For a time it was the largest settlement in the City of Washington, and the fragmentary accounts that survive portray it as well run but sometimes raucous. Hoban served as the on-site construction superintendent as well as the architect and had the added responsibility of managing all the laborers. Illegal alcohol sale, gambling, and vice were persistent problems, but on the whole Hoban proved to be an effective manager and these issues rarely got out of hand.

The nation's second president, John Adams, announced that he would move into the White House on November 1, 1800, a full month earlier than the date fixed by the Residence Act. Preparations for his arrival began early in the year, and the mansion was ready just in time.

To improve the grounds, Secretary of the Navy Benjamin Stoddert enlisted the help of Dr. William Thornton, the original architect of the U.S. Capitol. Stoddert wanted "something like a garden, on the north side," modeled after that of William Bingham, whose Philadelphia mansion featured some of the best-tended grounds in the nation. The great chore of grading the site would have to wait, but Stoddert wanted trees "planted at once, so as to make it an agreeable place to walk in." Thornton made a few sketches and consulted with City Surveyor Nicholas King, but it was soon clear that the task of planting trees and developing a pleasure garden on the grounds was not possible due to a lack of time, money, and manpower. The space remained in "great confusion," still full of temporary structures, unused building materials, and heaps of refuse. The ground itself was uneven and pockmarked by holes dug to gather clay for the brick kilns and other scars inflicted during construction.

Such was the sight that greeted President John Adams outside his new home. While he did not commit his opinion to paper, it could not have been favorable. First Lady Abigail Adams, who joined her husband two weeks later, complained to her daughter that the site lacked "the least fence, yard, or other convenience," but she also appreciated the "beautiful spot, capable

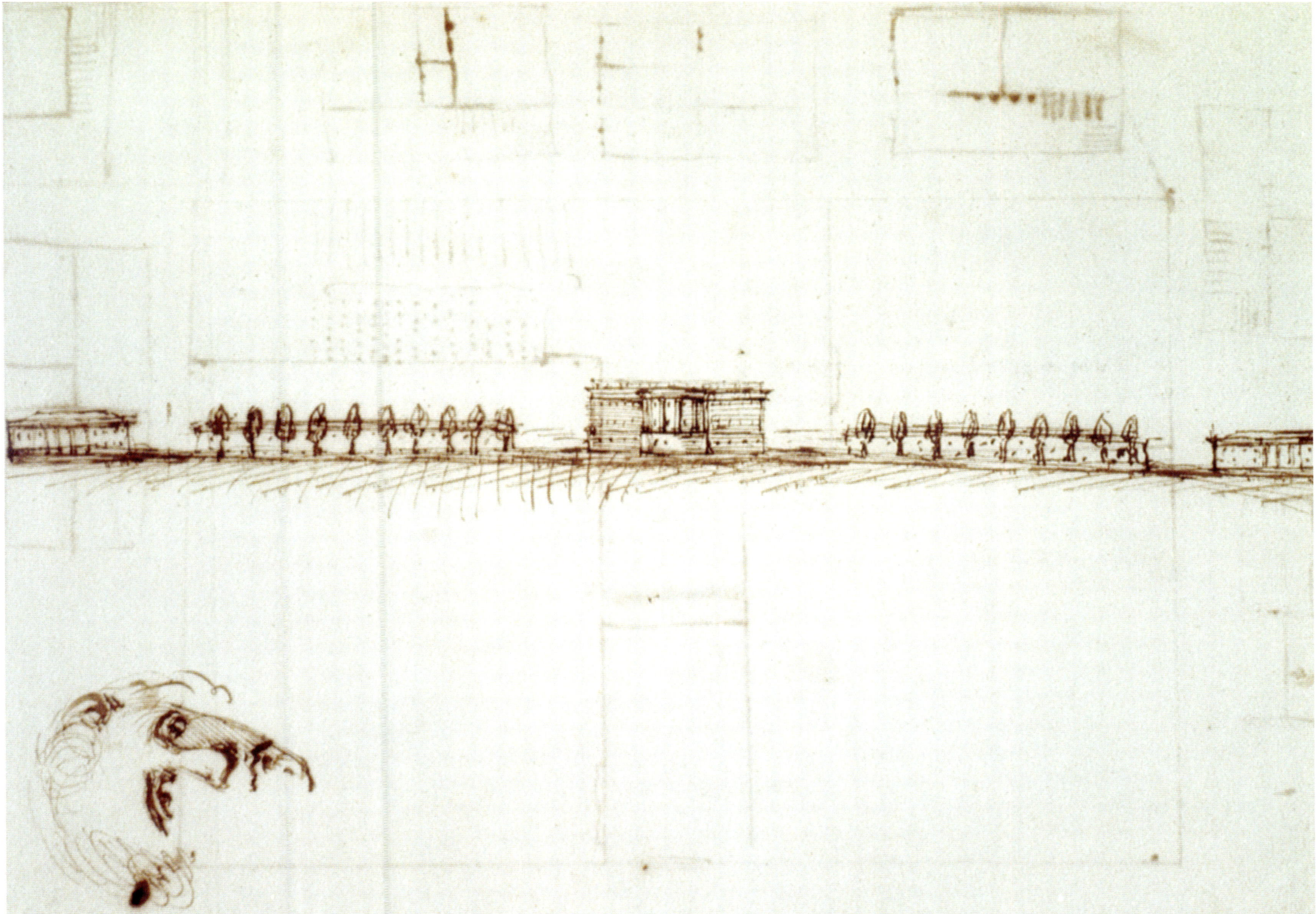

1347

TOM FREEMAN
©2000

President John Adams hosted the first New Year's Reception at the White House on January 1, 1801. The guests used a temporary set of wooden stairs on the south to enter the Oval Room. The north entrance to the White House was not yet complete. This painting, as imagined by artist Tom Freeman in 2001, illustrates the historic occasion.

of every improvement, and the more I view it, the more I am delighted with it." Her time in the mansion was very brief—less than four months. John Adams had narrowly lost his reelection campaign before she moved in, and so improvements to the grounds fell to the next president, Thomas Jefferson.

When Jefferson moved into the White House, in March 1801, the property still looked like a construction site. The brick kilns, now cold, remained in place, as did more than a dozen wood cottages. Hoban had begun dismantling the temporary settlement during the previous summer when he purchased ten of the cottages, rolled them away, and set them up as rental properties elsewhere in the city. He had little trouble finding boarders, as government employees began arriving, driving up the rents and making even the little cottages desirable. The government finally auctioned and removed the last cottages from the grounds in 1804.

Visitors described the White House Grounds in Jefferson's time as a "barren, stony, unfenced waste," "in a rough, wild state," littered with construction debris, piles of excess clay, and stumps of fine, old trees felled for firewood or perhaps for the floors of the President's House. An English gentleman wrote in his travelogue that after nightfall "instead of finding your way to the house you may perchance fall into a pit or stumble over a pile of rubbish." In his opinion, the sorry state of the grounds destroyed "every sentiment of pleasure that arises in the mind in viewing the residence of the president of a nation, and is a disgrace to the country." Jefferson agreed, and he began the process of cleaning up and developing the site. His dedication to the work transformed the landscape, so that by the end of his administration it was clear of debris and on its way to becoming the verdant greensward we know today. Although modified many times thereafter, Jefferson's vision became the basis for all future improvements and firmly established the interrelationship between the President's House and the President's Grounds.

Less than a week after his inauguration, Jefferson decided to replace the inconvenient outhouse with two proper indoor water closets. Construction of the indoor bathrooms was slow. The pipes for at least one of these facilities were not ordered until the fall of 1803, and so the privy probably remained an unwelcome eyesore but necessary part of the grounds until at least this time. During his first week in office Jefferson also authorized the hire of "carts and laborers to clear a passage, & level the ground, round the north side of the President's House." This statement, recorded by the city commissioners, seems to allude to the creation of the first north entranceway. Initially only a dirt path, this route led straight to the North Facade of the White House, thus making it the main, public side of the building, and it has remained so ever since.

In July 1801, Jefferson authorized the erection of a simple wooden post-and-rail fence that divided space within the huge plot of land once intended by L'Enfant to surround his aborted palace. This fence was more than 2,600 feet long and enclosed just under 9 acres. Inside the fence, Jefferson was vexed by the absence of smokehouses, stables, storehouses, and other ancillary structures essential to the day-to-day operation of even a modest estate. He knew that permanent outbuildings would clutter the grounds, but also realized that moving the required functions into the mansion was not an option. His solution was ingenious: two service wings appended to either side of the main block of the house.

These wings are now known as the East and West Terraces and comprise one-story, flat-roofed hyphens sunk into the natural north-south slope of the land. On the north, they barely stand above grade and are for the most part noticeable only because of the large lunette windows that provide light at ground level. On the south, Tuscan columns support a single-story rise from the level of the exposed basement of the mansion. The spaces between these columns were left open in the nineteenth century, allowing the domestic staff to come and go as they tended to their chores. The novel design provided workspace without seriously impinging upon the landscape. However, because the new wings were placed laterally across the crown of the ridge underlying the White House, they effectively divided the grounds into northern and southern sections along this line.

Construction began in May 1805, after Benjamin Henry Latrobe translated these initial renderings into working drawings. Jefferson had personally selected Latrobe, the finest architect and engineer in the nation, as first surveyor of the public buildings of the United States. In this position, Latrobe was responsible for all federal design and construction projects within Washington, D.C., and answered directly to the president. Consequently, Jefferson sought his advice and expertise in carrying out projects at the White House and on its grounds.

As originally conceived, the terraces would also have connected to the pair of government offices flanking the White House on the east and west. Completed about the same time as the mansion, these buildings housed the Treasury, State, War, and Navy Departments. It was logical to have the executive branch departments centrally located, but their addition to the landscape also became the first permanent intrusion onto the President's Grounds. Moreover, from an architectural perspective, the three buildings formed an odd ensemble: the red brick walls of the offices stood in stark contrast to the sandstone shell enveloping the presidential mansion. Jefferson intended that the terraces would create visual harmony by connecting the White House to the offices and also by filling in the space originally set aside for the much larger palace. No longer would the White House appear swallowed up by its surroundings or too small for its own grounds.

Unfortunately, this plan to link the White House with the executive offices failed in December 1806, when cold weather prevented mortar from setting properly and caused the leading segment of the East Terrace to collapse. At this point the terrace reached less than half its intended length. The accident became the talk of the town and embarrassed the president and his architect. Jefferson halted construction at the point of the collapse and shortened the West Terrace to match. The abbreviated work wrapped up in 1808.

Jefferson's larger plan for the White House Grounds called for a high wall of coarse stone to replace his earlier post-and-rail fence, and Latrobe envisioned that it would reach all the way out to the executive buildings and enclose up to 20 acres, roughly twice the area held within the existing wood fence. While most of the new wall was to rise above ground, to maintain the view down toward the Potomac River the southern section took the form of a sunken walled ditch known as a ha-ha. First employed about 1714 by English designer Charles Bridgeman, the ha-ha allowed for a physical barrier while creating the illusion of an unbroken landscape when looking out from atop the high ground. Relatively common in the Old World, the ha-ha was unusual in early nineteenth-century America, and its appearance on the President's Grounds probably stemmed directly from Jefferson's personal experience in landscape design.

In Jefferson's plan, a *tapis vert*, or green carpet, forms the central feature of the South Grounds, the same width as the main block of the White House and extending down to a steep declivity just inside the perimeter wall. Conspicuously, the landscape plan provides no details whatsoever concerning the treatment of the southwest section of the grounds. However, an associated topographical survey indicates that it was the most uneven part

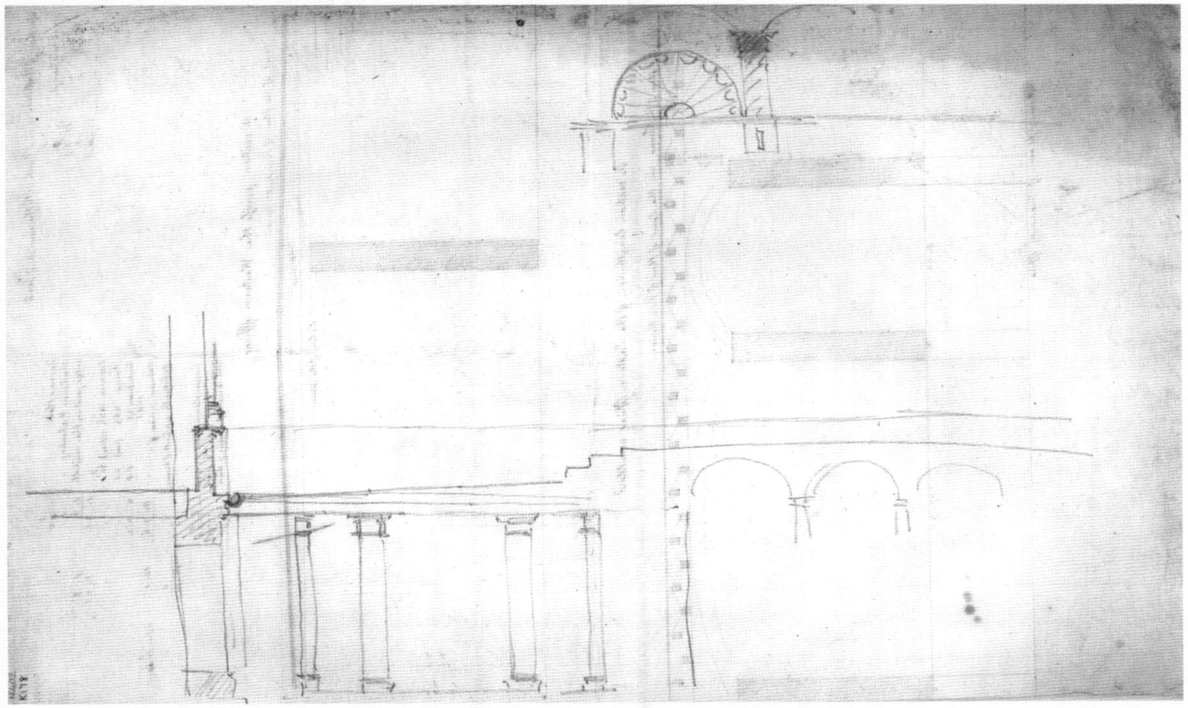

President Jefferson made this sketch for the East Terrace in 1803. The idea for terraces probably occurred to him in 1801, shortly after he moved into the White House, and by 1803 he resolved to have them built. He began sketching plans soon thereafter, inspired by a similar concept already used at Monticello, his Virginia home.

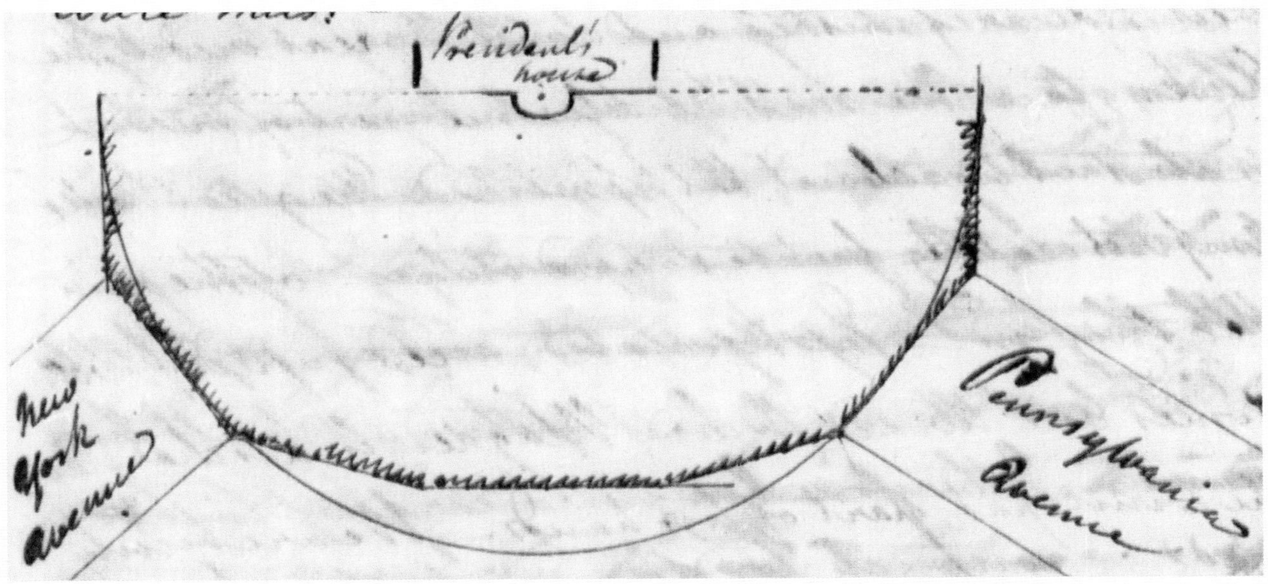

Benjamin Henry Latrobe sent this sketch to Thomas Jefferson on April 29, 1807, illustrating his intention to change the southern boundary of the President's Grounds from a semicircle to an oblong shape.

Jefferson's arch at the southeast entrance to the grounds is clearly seen in the background of this portrait of President Andrew Jackson, c. 1836–37, by Ralph E. W. Earl (left), and also in this period illustration of the view from the President's House up Pennsylvania Avenue to the unfinished Capitol (below).

of the entire site, as it had yielded much of the clay used for brickmaking. Jefferson probably intended that, once the grading and backfilling were complete, the west side should mirror the picturesque scene he envisioned for the east side.

The primary entrance onto the South Grounds was to be at the southeast corner, where the stone wall crossed Pennsylvania Avenue. The fourfold reduction in size between the President's Palace and the White House had pushed the mansion out of alignment with the avenue, leaving it without an architectural terminus. Jefferson now offered a solution in the form of a monumental triple-arched entrance gate for pedestrians and carriages. The landscape plan shows a driveway passing through the gate and then meandering past a garden, wood, and clump of shrubs and trees before continuing through to the North Grounds. Along the way, visitors would glimpse the White House through the foliage. Their little journey promised to be serene and picturesque.

The president planned a more formal treatment for the North Grounds. Broad, tree-lined allées would converge on the White House in a sunburst pattern. L'Enfant had intended for five major city thoroughfares, one street and four avenues, to meet in front of his palace, and the walkways now extended this axial arrangement toward the North Front of the mansion. The center radial, which followed the line of Sixteenth Street, also marked the entrance point onto the North Grounds. Through a gap in the wall and past a substantial pair of heavy stone piers, a gravel carriageway some 50 feet wide ran arrow-straight down to the front door of the White House, where a circular forecourt linked it with the winding drive coming up from the South Grounds.

Overall, Jefferson's landscape design was a hybrid of two philosophies. The first was the "natural" or English school of gardening, which tended toward the picturesque aesthetic of the South Grounds. He preferred this style and employed it at Monticello. The second was the French school seen in the North Grounds, which stressed clean lines, symmetry, and formalism. L'Enfant, the Frenchman, laid out the City of Washington as if it were one vast French-style garden. On the North Grounds, Jefferson was evidently content to follow L'Enfant's plan, but on the South Grounds, he let his own tastes and sensibilities shine through.

It remains unclear just how much of Jefferson's ambitious landscape plan was ever realized. What is known for certain is that the abbreviated East and West Terraces were built and the carriageway on the north side was laid down. By the time Jefferson left office, a new set of stone stairs reached up to the front door and were supported by a massive groin vault. Latrobe began construction of the stone wall of enclosure in April 1807 and finished in September 1808. Its height varied between 7 feet, 6 inches, and 8 feet, depending on the underlying grade of the ground. The grand entrance gate at the southeast corner of the President's Grounds was erected, although little information has survived concerning the process of its design and construction. Latrobe usually gets the credit, but Jefferson was surely involved as well. This monumental structure, 25 feet high and up to 66 feet long, served as an admirable surrogate for the White House in anchoring that end of Pennsylvania Avenue. Late in his life Edmund Baker, Jefferson's manager at Monticello for twenty years, recalled: "When I was there, the President's house was surrounded by a high rock wall, and there was an iron gate immediately in front of it, and from that gate to Capitol, the street was just as straight as a gun barrel."

Latrobe also supervised the grading of the President's Grounds, but the magnitude of the job made completion impossible before the end of Jefferson's administration, especially as the appropriations authorized by Congress dwindled after Latrobe overspent. The Herculean task stretched on for decades, with teams of laborers engaged in earth-moving projects as late as 1834.

Jefferson was a life-long gardener. At Monticello he experimented with ornamental and productive plant cultivation. While it is romantic to imagine him in a work tunic, shovel in hand, planting trees, flowers, and vegetables on the White House Grounds, there is little evidence that he gardened there. Apparently he did send for two loads of aspens from Monticello, but where, or if, they were planted is unknown. Yet although he contributed few, if any, plants to the landscape, he set the stage upon which all subsequent presidents have acted. His successor, President James Madison, planted trees and vegetables in abundance, and Jefferson may even have placed the requisite orders before retiring. In 1818, Architect of the Capitol Charles Bulfinch created a plan for continuing improvement of the landscape. Though the plan does not survive, Bulfinch assuredly looked to Jefferson's earlier concept for inspiration and guidance. It is clear that Jefferson's landscape plan and his overarching vision transformed the vague idea of a White House garden into a reality and served as the basis for all future development. Where others saw only a barren, abused patch of earth, Jefferson possessed the skill and imagination to envision its potential to become an elegant and functional space befitting the office of the president of the United States of America.

This design for a tapis vert *(an unbroken stretch of land)* is attributed to Thomas Jefferson (right). Filed in the collections of the Massachusetts Historical Society as a miscellaneous garden plan for an undetermined location, it depicts a rectangular panel with an interior arrangement of graveled walks, grass islands, hedgerows, and espalier frames for vines or fruit trees, all surrounded by a 4-foot wide border of flowers. This flower border is key, and suggests that the sketch may be a detail drawing for the South Grounds of the White House, as Jefferson's c. 1807 landscape plan (below) includes a long rectangular area immediately south of the mansion. Faded, minuscule script denotes that this space was to be bordered by flowers, precisely as indicated in the drawing at right. Moreover, the panel's dimensions match those of the rectangle on the South Lawn almost perfectly, adding further weight to the idea that the heretofor unidentified drawing is in fact a long-overlooked aspect of Jefferson's grand vision for the President's Grounds.

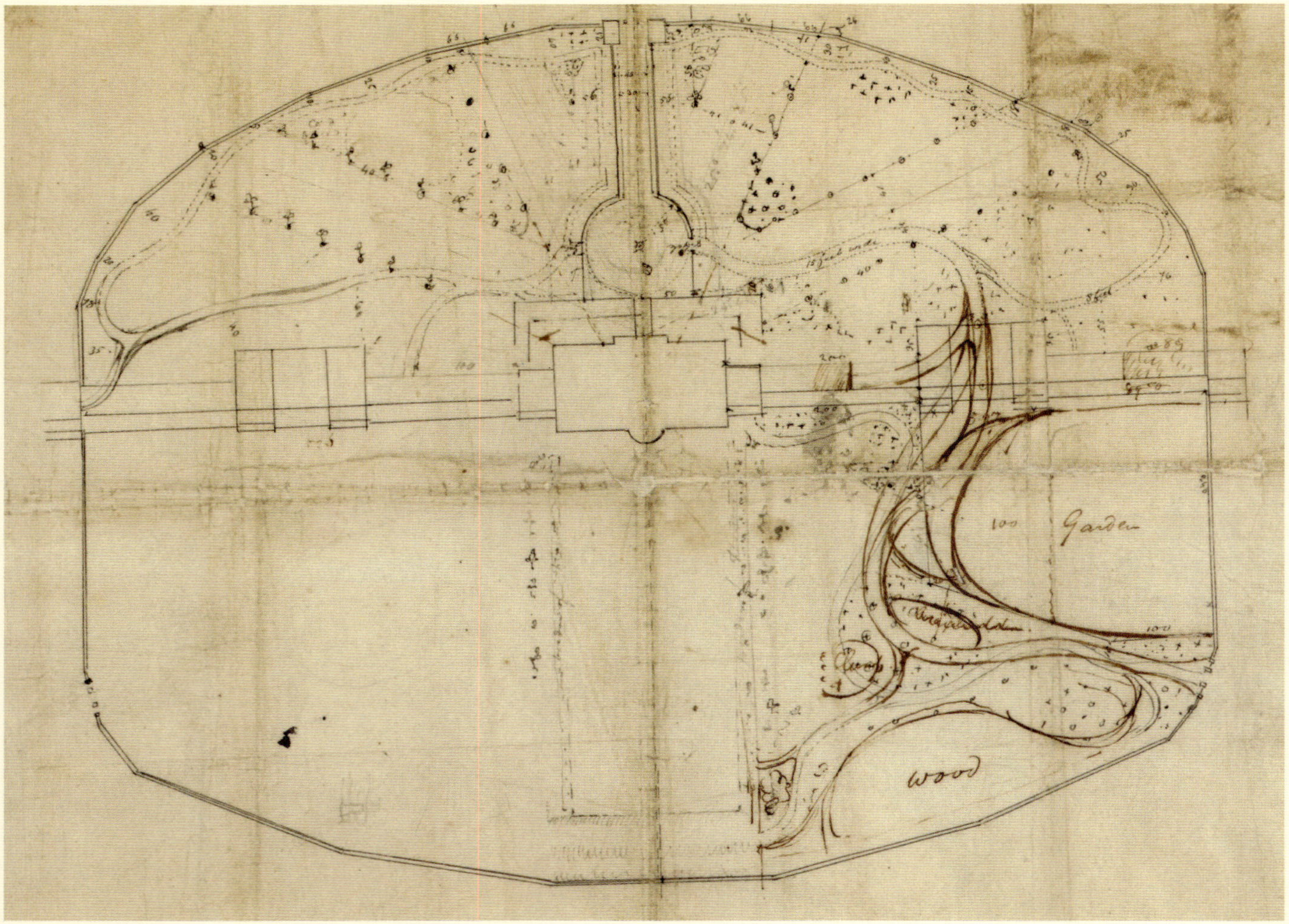

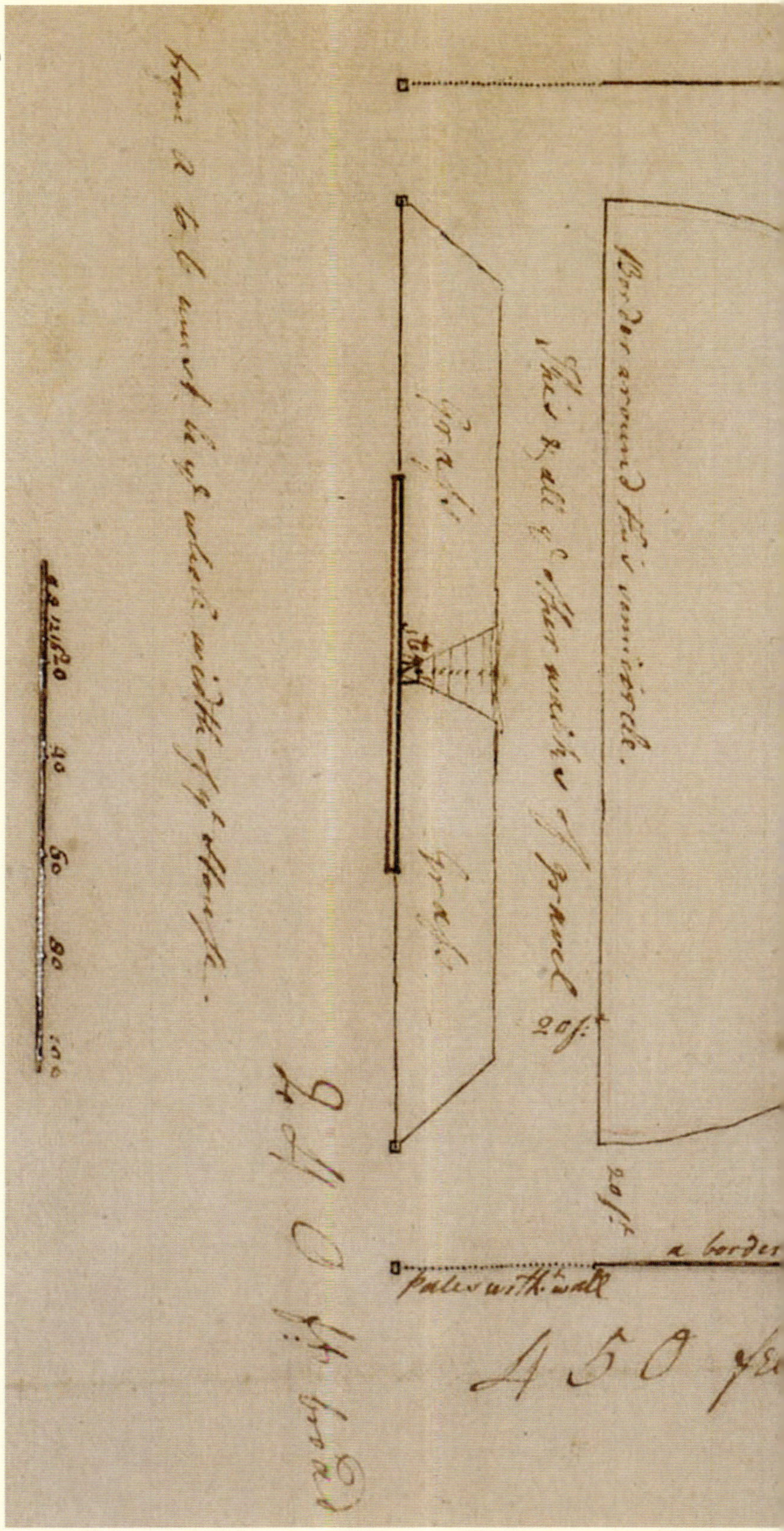

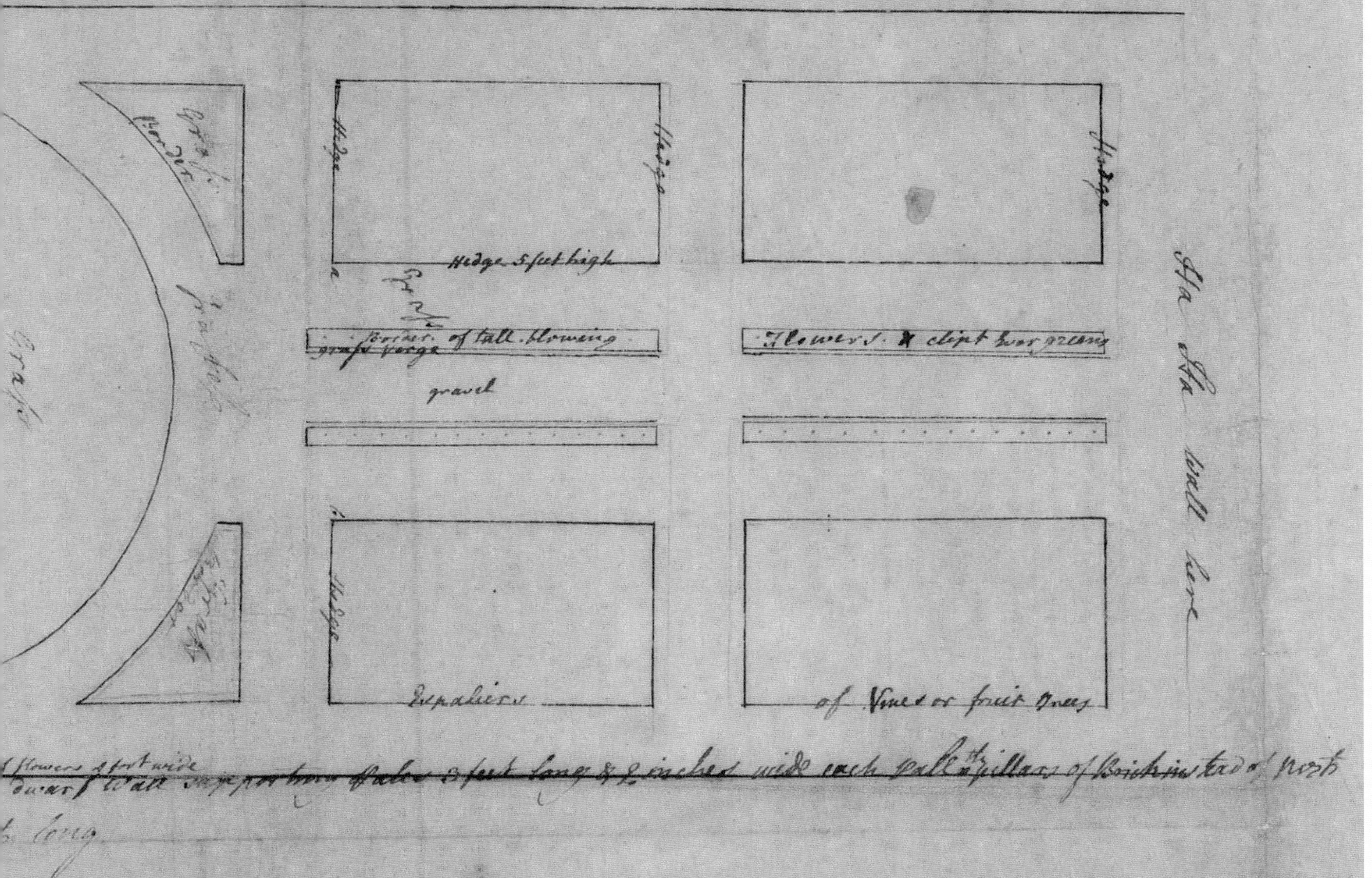

Grass Border
Hedge
Hedge
Hedge
Hedge 5 feet high
Grass
Border of tall flowering
grass verge
Flowers & clipt Ever greens
gravel
Ha Ha wall here
Grass
Hedge
Espaliers
of Vines or fruit trees
Flowers 5 feet wide

In 1850, President Millard Fillmore hired Andrew Jackson Downing to make a new landscape plan for all of the "Public Grounds" of Washington, D.C., including the President's Grounds. Downing's draft plan, which he submitted to the president the following year, was a masterpiece. In most places the plan is flush with detail, but the White House site is rendered in only a rudimentary way. Downing tragically died in a steamboat accident before he provided additional drawings, and his plans for the presidential enclave died with him.

Downing's death stands out as the most "what if?" moment in the history of the President's Grounds. Had he lived, the landscape may have been transformed to coincide with the grandiose, Victorian aesthetic that governed his plan for the city's Public Grounds. As with Jefferson before him, Downing had the opportunity to integrate all parts of the presidential landscape into a cohesive whole. Instead, eight decades of piecemeal, ad hoc changes followed his death. At the White House change is inevitable, but there was no master plan to work from. Over the years design after design, ranging from the reasonable to the ridiculous, was rejected in turn, usually because of a lack of funding. In the absence of a master plan, the grounds evolved to suit the needs of successive presidents. Consequently, the landscape they enjoyed, while pleasant, did not display a unified appearance.

The next president to take a serious interest in the cohesive development of the President's Grounds was Franklin Delano Roosevelt. He had never been a gardener, suffering from a lifetime of allergies and the loss of the use of his legs at midlife from polio. Nonetheless, he recognized the problems in the grounds, and he wanted them solved. As a first step, he retained Frederick Law Olmsted Jr., a renowned landscape architect who had served as an adviser to architect Charles Follen McKim during Theodore Roosevelt's renovation of the mansion in 1902. Olmsted later recalled that McKim had been "in principle completely sympathetic with my views" toward the landscape but fixing the "outrageously bad functional and artistic condition" of the White House left no time or money "to grapple broadly and completely with the problem of the grounds as a whole." Now, thirty years later, another Roosevelt was giving him a second chance to "grapple" with the long-standing issues.

Despite its haphazard development, both the president and landscape architect respected the historical significance of the President's Grounds. Olmsted hired Professor Morley Jeffers Williams of Harvard University's School of Landscape Architecture, who was then restoring George Washington's gardens at Mount Vernon, to research the subject. Olmsted felt that understanding the landscape's history before he advised on specific changes would be essential to fulfilling Roosevelt's charge. The need for such background study seems obvious today, but at the time it was a novel concept. The importance of America's historic sites was only then beginning to be realized, and the government had never undertaken any major historical study of the White House or its grounds. Williams's task was therefore unique, and his findings, although not particularly detailed, have served as the basis for every subsequent study of the White House landscape, including this one.

While Williams sorted through old drawings and long-forgotten reports, Olmsted toured the grounds and took notes. The giant, venerable trees enthralled him, but he puzzled over the discordant elements left behind by previous periods of development. He deemed the overall effect to be "distinctly respectable" but also "distinctly disappointing." The White House was a special building; its grounds deserved to be equally special. More than anything else, President Roosevelt wanted the site to be greater than the sum of its parts. It needed to be governed by some sort of overall concept. Olmsted submitted his recommendations on October 24, 1935; the Olmsted Plan remains the baseline document for the use, maintenance, and development of the President's Grounds to this day.

Olmsted viewed the landscape holistically and recognized that the White House itself provided a clear dividing line between the North and South Grounds. The South Grounds took on a park-like appearance, with thick groves of trees flanking a broad open vista that sloped down toward the Washington Monument and, after its completion in 1943, the Jefferson Memorial in the far distance. The return view was equally important, as it provided the public with a clear look back up at the South Front of the mansion. All flowerbeds, shrubs, trees, and other obstructions along this sightline would, according to the Olmsted Plan, be removed, and the formal garden areas restricted to the immediate vicinity of the White House. Automobile parking was banned, and the driveways that crossed the grassy expanse sunk down into the ground so as not to intrude on the view.

The North Grounds, which since Jefferson's time had been given over to a more formal design, continued in this vein. Some of the foliage needed to be thinned out, but the stately rows of oaks and elms remained to shade the lawn. Many of these trees had been planted in the nineteenth century, and Olmsted wanted them preserved. He also made sure to retain a clear view of the iconic North Facade of the White House, which could be seen from a much closer proximity than that on the south.

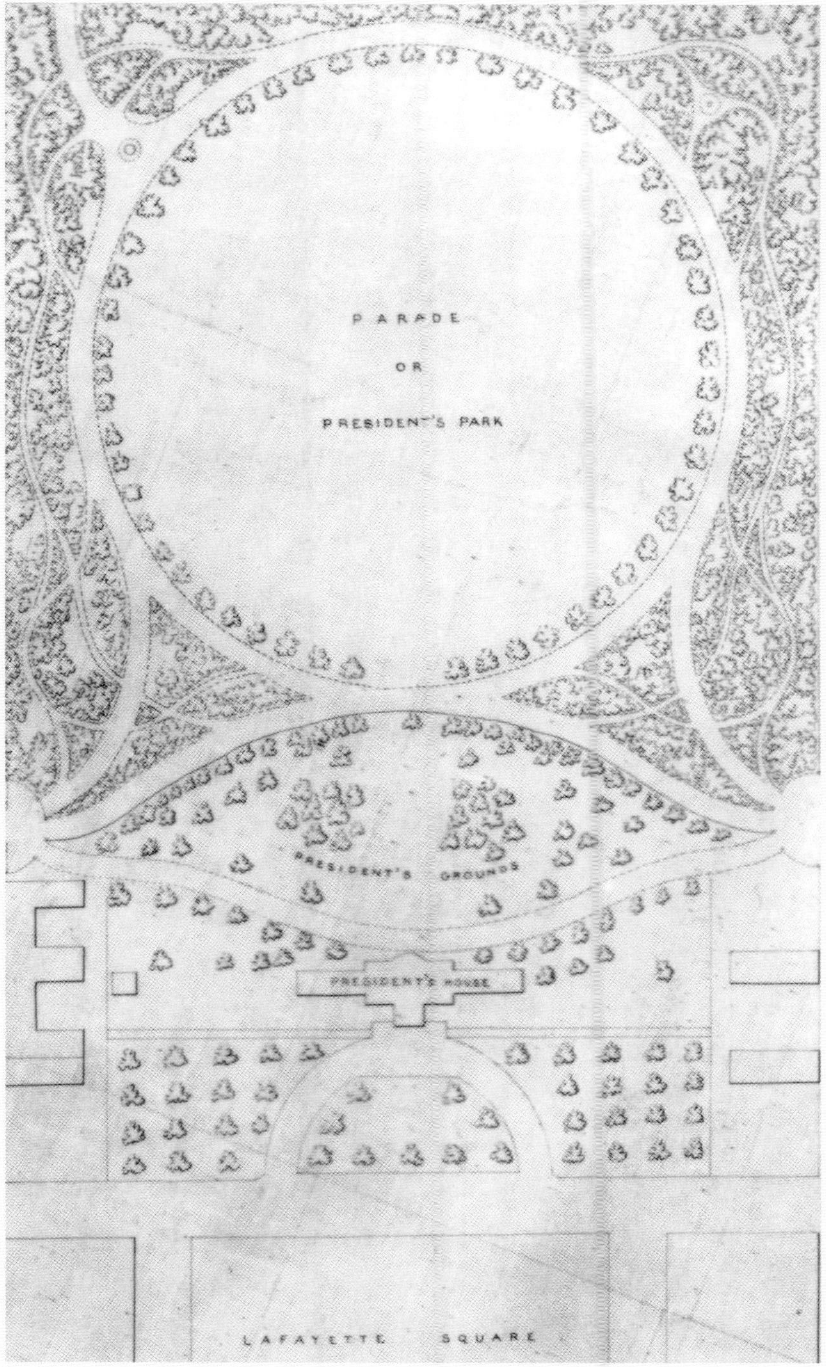

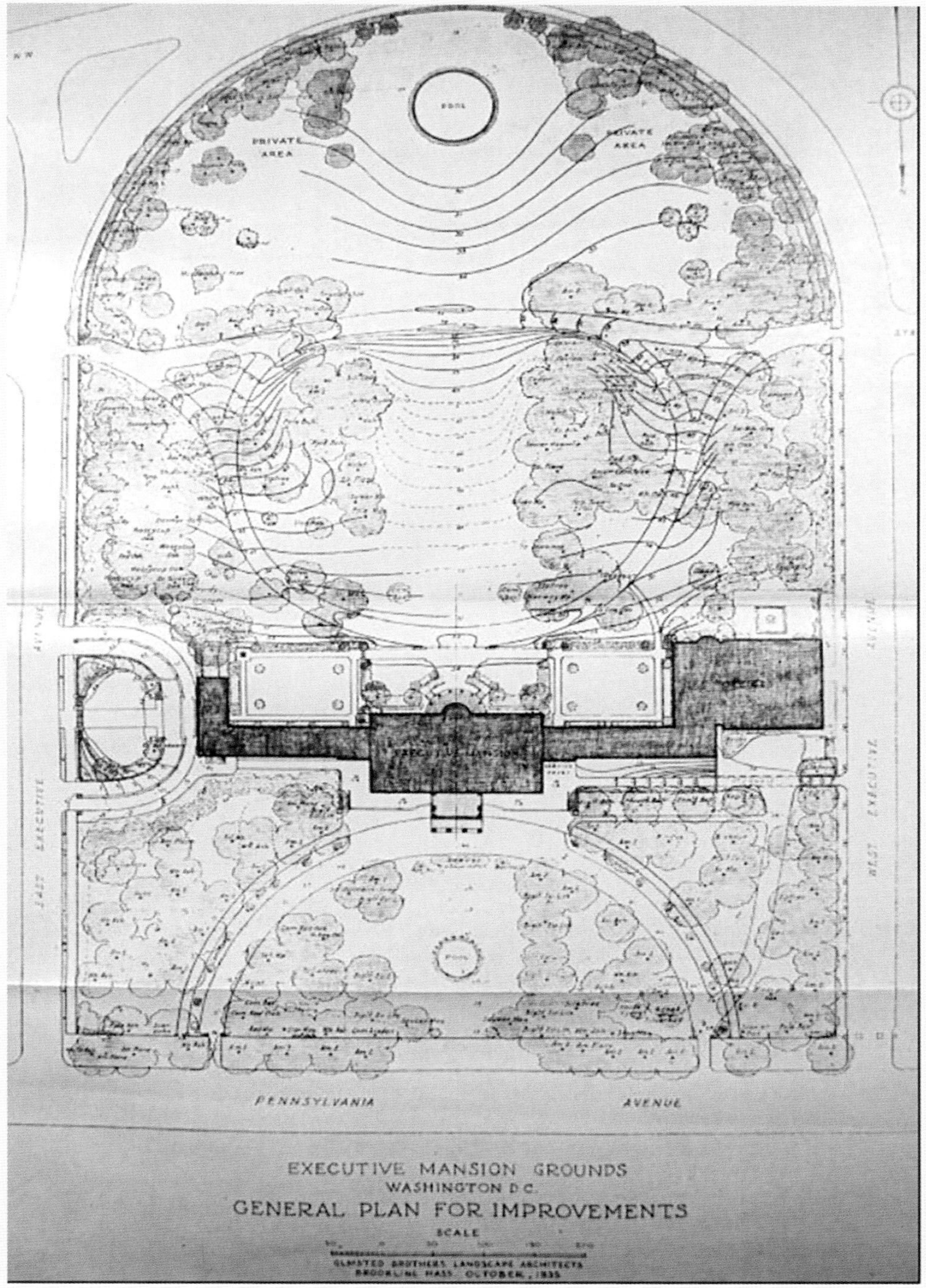

Two plans that played significant roles in the evolution of the White House landscape, the Downing Plan, 1851 (far left) and the Olmsted Plan, 1935 (left).

The National Park Service maintains the White House Grounds and President's Park. They are responsible for beautiful seasonal plantings including the fall chrysanthemums seen here.

There were, of course, additional details and further recommendations, but part of Olmsted's genius was the fact that he chose to paint his vision for the President's Grounds with broad brushstrokes. He looked at the big picture and generally did not get bogged down in minutiae. He was also smart enough to realize that the landscape would need to continue to evolve according to the particular needs and desires of each first family. His plan is therefore flexible and not unduly restrictive. Most changes, be they as minor as planting a new tree or as major as installing a swimming pool, can be accomplished without impinging upon the character of the landscape as a whole. All the presidents since Franklin Roosevelt have supported Olmsted's ideas, allowing the White House Grounds to achieve a dignified and endearing quality bestowed only by the slow passage of time.

Today the President's Grounds comprise approximately 18.7 acres and form the central component of a larger 88.2 acre landscape known as the President's Park that also includes Lafayette Park, Treasury Building Grounds, Eisenhower Executive Office Building Grounds, Ellipse, Sherman Plaza, and the First Division Monument. The current boundaries of President's Park match those reserved as the location of the President's Palace, as shown in the initial 1791–92 plans for Washington, D.C.

The President's Grounds are managed by the Office of the White House Liaison for the National Park Service, and it is in honor of the Park Service's one-hundredth anniversary that this book is being published at this time. The President's Park, and all national capital parks, were placed under the jurisdiction of the National Park Service in 1933. At the beginning, in 1791, the three commissioners of the District of Columbia had responsibility for this land. Jefferson briefly abolished the office of the commissioners, but in 1816 Congress restored it, and in 1849 it was placed under the Department of the Interior until 1867, when the Army Corps of Engineers, Office of Public Buildings and Grounds, took over. In 1925 Congress established the Office of Public Buildings and Public Parks of the National Capital, which had responsibility for the President's Grounds until the reorganization of 1933, which moved it back into the Department of the Interior.

Daily and seasonal maintenance of the President's Grounds is currently carried out by landscape architects, horticulturalists, and gardeners. It is to them and to the National Park Service that the president and the public alike are indebted for the well-manicured green spaces, magnificent trees, and many gardens full of flowers that lift the spirit and rest the soul in the heart of the nation's busy capital.

These illustrations, made by artist Rodica Prato c. 2011, include the gardens, trees, drives, fencing of the North and South Grounds.

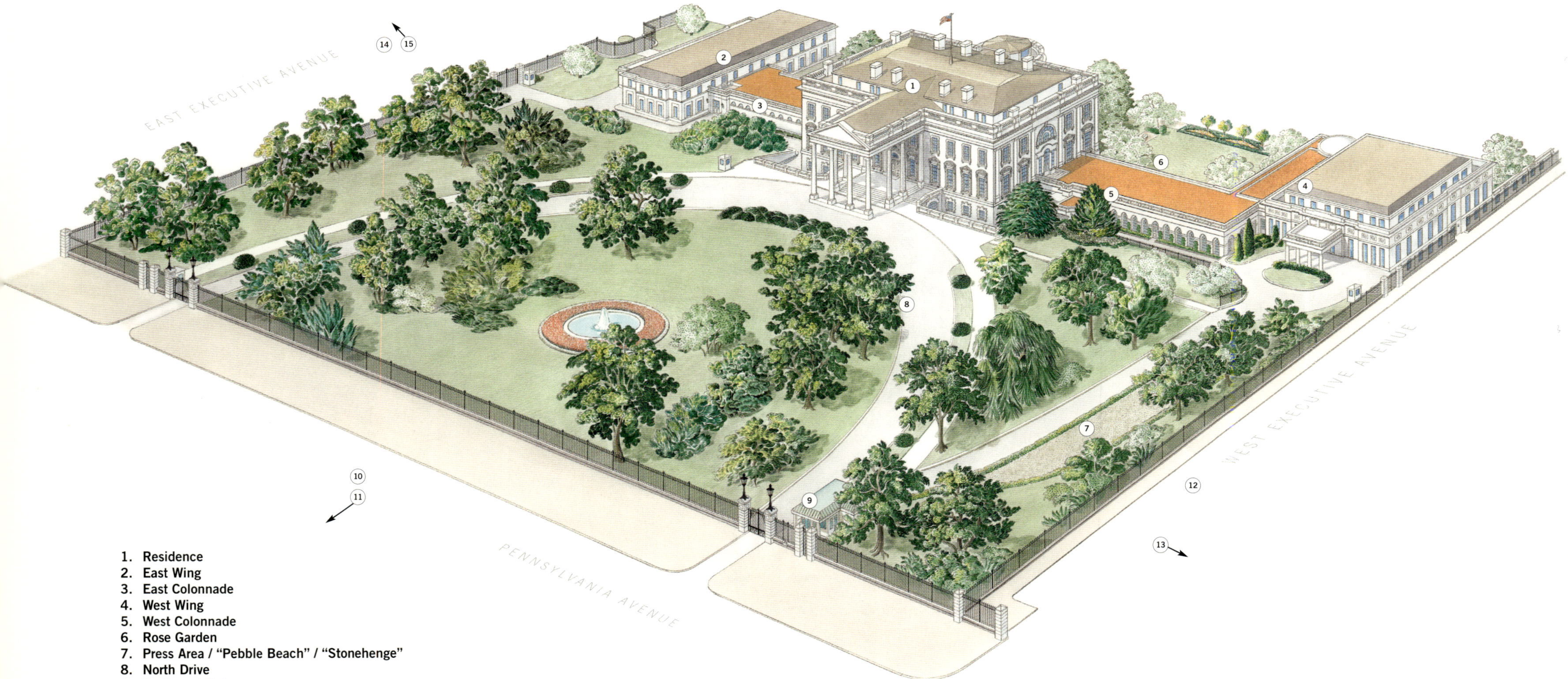

1. Residence
2. East Wing
3. East Colonnade
4. West Wing
5. West Colonnade
6. Rose Garden
7. Press Area / "Pebble Beach" / "Stonehenge"
8. North Drive
9. Northwest Gate
10. Pennsylvania Avenue
11. To Lafayette Park
12. West Executive Avenue
13. To the Eisenhower Executive Office Buildling
14. East Executive Avenue
15. To the Treasury Building

1. Residence
2. East Wing
3. East Colonnade
4. West Colonnade
5. West Wing
6. East Garden/Jacqueline Kennedy Garden
7. Rose Garden
8. Swimming Pool
9. Tennis Court
10. Children's Garden
11. Kitchen Garden
12. East Entrance
13. E Street
14. To the Ellipse
15. Pennsylvania Avenue
16. To Lafayette Park
17. West Executive Avenue
18. To the Eisenhower Executive Office Buildling
19. East Executive Avenue
20. To the Treasury Building

PART I
THE GROUNDS

THE NORTH GROUNDS *26* / THE SOUTH GROUNDS *44* /
THE WEST GARDENS *76* / THE EAST GARDENS *98* /
VISITING THE WHITE HOUSE *110*

THE NORTH GROUNDS

The Front Lawn of the White House

THE WHITE HOUSE, official residence of the president of the United States, is also affectionately seen as the people's house, and its grounds, the people's grounds. This duality of possession was apparent from the beginning, when President Thomas Jefferson first weighed the merits of presidential privacy against the right of any citizen to visit the White House and stroll upon its green grass. Today, safety and security concerns justifiably limit access, but two hundred years ago this was not the case. Jefferson himself was known to receive visitors at almost any hour of the day or night, when those calling might encounter him in slippers and robe. Jefferson presented himself as a true republican citizen, but he knew that the head of state required some ceremony. Ever the apt politician, he devised a compromise. The parlors on the State Floor, except for the president's office, would be

In the case of the White House, the front lawn is seen from Pennsylvania Avenue, and every year countless thousands pose for photographs here, ensuring that this north view remains its unquestionable public face.

This stereograph from the 1860s captures pedestrians pausing on the sidewalk in front of the White House.

made accessible more or less for the visiting public, while the bedrooms and sitting rooms upstairs would be private quarters, off limits except by invitation. The overall arrangement was in keeping with the traditional design of an ordinary home where the living room, dining room, and entertainment and functional spaces are separated from the bedrooms, dressing rooms, and similar areas reserved for the family's exclusive use.

This differentiation of space extended to the landscape. Jefferson envisioned the North Grounds as public space, where business callers and tourists entered, and intended for the South Grounds to be private. This balancing act of public access and presidential privacy remained problematic for more than a century, as visitors increasingly gained access to areas that had previously been off limits. Nonetheless, the functional distinction between the North Grounds and South Grounds remained and served as the basis for the subsequent development of the White House landscape.

The North Grounds thus became the president's front lawn, and as with most American homes it provides a first—hopefully favorable—impression to the passerby. Today, the view remains open, pleasant, and neat. The formal and symmetrical landscape treatment enhances the neoclassical elegance of the White House itself, accentuating and showcasing the home of the president, in the same way a more modest front lawn improves the curb appeal of a typical American home.

The North Entrance: Fence, Gates, Piers, and Driveway

THE SUBSTANTIAL CURB appeal of the president's front lawn is increased by the skillful arrangement of iron fences and gates, stone piers, and the wide-arcing driveway that together constitute the principal entrance to the grounds. These features frame and anchor the house within the landscape, drawing our eyes straight ahead toward the most famous front door in America. Functionally accessible and secure by design, the North Entrance is also highly ornamented, so it greatly enhances the overall character of the North Grounds. Moreover, these grounds hold some of the oldest extant landscape elements present anywhere on the President's Grounds.

President Monroe's 1818 entrance replaced a portion of President Jefferson's stone wall located directly in front of the White House. It is unclear whether the old wall was completely removed or cut down to a height of 1–2 feet and then used as a base course for anchoring Paulus Hedl's iron railing. If it was cut, then a short knee-wall is all that remains of a once towering edifice that was torn down piecemeal in the coming decades. There is no evidence to suggest that the large eagle statues seen in this c. 1811 watercolor by B. Henry Latrobe were ever actually installed.

This hand-colored print entitled The President's House, *engraved and published by Henry and James Stone in 1826, shows the North Entrance begun in 1818, which included two large double gates for carriages and two smaller single gates for pedestrians. The ironwork was created by Paulus Hedl of Westchester County, New York.*

Construction of the North Entrance began in 1818, during the presidency of James Monroe, when New York ironworker Paulus Hedl received the contract to supply the White House with four new gates: two large double-gates for carriages, two smaller single-gates for pedestrians, and 180 feet of iron railing to run between them. Each set of gates was to be flanked by a set of stone gate piers of appropriate size, and three glass oil lamps and wrought-iron brackets were staggered along the railing to provide nighttime illumination.

Continuing what Monroe began, President Andrew Jackson widened the North Entrance in the 1830s by moving the gates and piers farther east and west and adding additional railing matching Hedl's. By the 1870s, this Hedl-style fence spanned the entire width of the North Grounds, creating one continuous run along Pennsylvania Avenue.

Hedl was a master of his craft; his fence and gates proved strong, elegant, and highly refined. His original gates served the president until 1976, when security called for replicas of stronger, reinforced metals. The wrought-iron fence, how-

PRESIDENT'S HOUSE.

Published by H. & J. Stone, Washington, 1826.

Hedl's iron gates were captured by artist Thure de Thulstrup in this wood engraving titled Tragedy at Washington—The Night-Watch Before the Executive Mansion, *which recorded the scene outside the White House as President James Garfield slowly neared death from an assassin's bullet in the summer of 1881.*

In 2002, an original set of nineteenth-century White House gates were discovered on the grounds of River Farm, the headquarters of the American Horticultural Society in Alexandria, Virginia. Evidently forgotten for decades, the ironwork was rusty, bent, and in need of repair. The society recognized the significance of the gates and in 2005 had them restored and installed in its garden on public display.

ever, remains in place along Pennsylvania Avenue to this day, its posts further stabilized by iron braces inside the fence and embedded in concrete at the base. The individual palings are capped with spearhead finials, and the posts with cast-iron finials reminiscent of a classical urn. Of the three lamp stands incorporated into the original design, two remain in place, one at the very center of the fence line and the other to the east. Both the original and modern gates were designed to match the fence, incorporating similar finials.

In 1852, during the presidency of Millard Fillmore, four large iron gas lamps were installed atop each of the carriage gate piers. Records show that this ironwork came from the Philadelphia foundry of Robert Wood and Company, one of the premier makers of ornamental ironwork in mid-nineteenth-century America. Period accounts describe the lamps as "massive candelabras," and as they are still in place, it is easy to see that this was no exaggeration. At the time of their installation, each lantern utilized five separate burners, but in the twentieth century they were electrified and continue to illuminate the North Grounds today.

The replacement gates of 1976 are of the same design as the original iron gates now on display at River Farm. Across the bottom of the gates is a series of overlapping iron arches inset with trefoils. This whole is surmounted by a horizontal band of a simplified Greek key, above which are palings of the same design as the fence. Finally, the upper right and left spandrels of the gates are decorated with mirror-image fans of scrolled ironwork that form a large round arch when the gates are closed.

This iron lamp on the Northwest Gate, pictured here in 2012, is one of four installed in 1852 that remain in place today. Originally gas, they were electrified in the early twentieth century. One of the laterns is inscribed with "Jos. Neumann / Phila., Pa."

The individual palings on Hedl's gate are capped with heart-shaped finial spearheads with splayed curves on both sides. The posts are topped by an elaborate cast-iron finial reminiscent of a classical urn, possibly based on the Greek cup form known as a kylix. *The finial is composed of an acanthus leaf base surmounted by a rosette. A scalloped urn or sword guard sits at the top and is capped by a decorative splayed curve and rosette.*

As devised by Hedl, the entire fence has an integrel iron bar across its base, which rests on a stone plinth wall capped by an Aquia stone cap. A similar iron bar stabilizes the entire fence across the top beneath the finials.

Paulus Hedl's old ornamental cast iron lamp bracket is seen at the center of this photograph, which captures workers installing sharp metal spikes on the fence on July 1, 2015. The spikes were added in an effort to prevent intruders from jumping the fence after several such incidents in less than a year.

This 1833 drawing made during Andrew Jackson's presidency depicts the decorative anthemion iron railing along the North Facade, which would also soon replace the old post-and-chain fence along one side of the semicircular drive.

The final component of the North Entrance, the semicircular driveway running between the iron gates, amounted to little more than a dirt track in its early days. Although designed as a grand, sweeping approach to the North Front of the White House suitable for use by the president, diplomats, and honored guests, it was originally left ungraveled as a cost-saving measure. The results were most unfortunate. Horse hooves and carriage wheels kicked up a choking cloud of dust in dry weather, while rain turned the driveway into a muddy quagmire. These problems persisted until the fall of 1834, when funds were finally requisitioned and workmen began hauling cartloads of gravel onto the North Grounds and spreading it over the driveway. With the construction of the North Portico in 1829, the driveway now passed beneath its sheltered overhang, allowing passengers to come and go without having to step out into foul weather.

A simple post-and-chain fence originally lined the semicircle to protect the North Grounds from "injury by carriages, horses, &c." In the 1830s, this utilitarian fence was replaced by a low, heavy iron railing featuring a decorative anthemion pattern characteristic of the Greek Revival style popular in the United States at the time. Matching runs of anthemion railing were also installed along the North Facade of the White House. The designer was most likely Paulus Hedl, who furnished a similar railing for the east side of the U.S. Capitol in 1820. The anthemion railing remained along the North Drive until 1873, when it was replaced with a short granite curb. That bordering the mansion remained a bit longer, until 1902.

Around the mid-nineteenth century, pairs of carriage gates, stone gate piers, and iron candelabras were installed east and west of the North Portico. Combined with a high iron fence added along the outer edge of the North Drive, the features controlled movement between the North and South Grounds of the White House. Most of this ensemble was taken down in the early 1870s, and only the southern gate piers remain in place today.

Photographed from the northeast during the Lincoln administration, this stereograph captures the view up the broad stone sidewalk beside the graveled roadbed of the North Drive. It shows the anthemion railings installed in the 1830s (sketch at left) as well as the taller iron fence, gates, gate piers, and lamps added around midcentury.

The Thomas Jefferson Statue: First Centerpiece of the North Grounds

THE NORTH ENTRANCE received an additional ornament in 1847 with the installation of a bronze statue of Thomas Jefferson in front of the North Portico, in the center of the grassy plot surrounded by the semicircular driveway. Sculpted in 1833 by the celebrated French artist Pierre-Jean David, more commonly known as David d'Angers after his place of birth, the statue was cast in Paris by Honoré Gonon and his two sons under a commission from Uriah P. Levy, a career officer in the U.S. Navy. The following year, Levy purchased Monticello, which had suffered serious neglect since Jefferson's death in 1826. Rather than installing the statue at Monticello,

By the time Mathew Brady took this photograph of the North Front of the White House in 1861, the Jefferson statue had been in place for almost fifteen years and was in poor condition. Period newspaper reports described it as being "covered with verdigris which is fast eating away the lines of his face, and the roll of parchment in his hand," and rather ingloriously dubbed it "a nondescript bronze statue, purple with years and neglect."

Levy chose to donate it to the United States, but Congress could not come to a consensus on whether to accept the gift. Jefferson, a Democratic-Republican, proved one of the nation's first polarizing political leaders, and members of the rival Whig Party did not want to accept the statue, especially as no president had as yet been thus immortalized in Washington, D.C., not even Washington himself. It apparently made little difference that the bronze actually depicts Jefferson not as president but as author of the Declaration of Independence, scroll and quill in hand. After much debate, Congress simply let the matter drop and the Jefferson statue remained in limbo for the next thirteen years, uncertainly displayed within the Capitol Rotunda because no one could agree on what else to do with it.

Enter President James K. Polk, a Jacksonian Democrat and self-styled "Jeffersonian" politician. In 1847, Polk happily called for the statue and had it set up on a pedestal of stuccoed brick in front of the White House. Controversy followed the statue to its new home, and because of the nature of American party politics, it became the first and only statue of a president ever erected on the White House Grounds. Twenty years later, the Jefferson statue remained the centerpiece of the North Grounds, but it showed signs of serious deterioration, possibly exacerbated by deferred maintenance during the Civil War. Congress even admitted that the statue was in poor condition, going so far as to decry it as "an object of neglect and reproach." To their credit, the members of Congress put aside political differences and ordered the statue to be placed under cover, thoroughly cleaned, and repaired.

By February 1874, the statue and its brick pedestal had been relocated to the east side of the White House, evidently in anticipation of carrying out the repairs out of public view, but upon closer examination, the damage to the statue was found to be worse than expected. The government shipped the statue to Robert Wood's foundry in Philadelphia, which completed the repairs and returned it to Washington by June 1875. Meanwhile, the ownership of the statue came into question. Because of the previous congressional indecision, the bronze legally remained Levy's property until his death in 1862. Now his brother lobbied Congress to either accept the gift on behalf of the people of the United States or return it to the family. If accepted, he insisted that it be installed in the U.S. Capitol, where he felt it ought to have been located all along, so it would not be exposed to the elements or ever again suffer so greatly for want of regular care. This time around, Congress assented and so the statue of Thomas Jefferson never returned to the President's Grounds. It remains in the Capitol to this day.

The North Fountain: Millions of Drops of Water

AFTER THE STATUE OF THOMAS JEFFERSON was removed, a decorative fountain was installed by President Ulysses S. Grant as the central object of honor on the North Grounds. As originally constructed, this gravity-fueled fountain comprised a small circular bowl of brick and cement with a single water jet rising from its center. A few years later the jet was replaced with a "French gurg" fitting that forced the pressurized water through five hundred small nozzles, creating a dazzling, delicate web of little streams. In 1883, a new, larger, 40-foot diameter stone-coped basin replaced the original so that the water pressure could be dialed up to full-throttle without the streams shooting past the edge of the bowl.

In 1877, a White House visitor described the fountain recently fitted with a new "French gurg" spray as seen in this stereograph: "The water springs in a multitude of jets, from a cluster of mock plants, twelve or fifteen feet into the air, and resolves itself into millions of drops which fall in a swift splashing rain into the scalloped basin. The center jet is heavier and higher, and in the moonlight or starlight I have seen it flashing like the gleam of a shining blade, when the rest of the fountain was quite invisible."

John Stewart's 1893 plan for the North Lawn features an elaborate pinwheel-shaped parterre banded with flowers surrounding the fountain. Inside the parterre, additional flowerbeds shaped like horseshoes, half-moons, circles, and stars, continued the eye-catching display. Petunias, lady-slippers, old maids, and phlox predominated during the summer, giving way to scarlet geraniums, salvia, and the ash gray leaves of the dusty miller plant in the fall. At the center of it all stood the fountain itself, with six graveled walks radiating outward and cutting through the plantings. Two long, narrow beds and two stars were also located just outside the main arrangement, flanking a short walk that connected with the North Entrance drive directly opposite the North Portico.

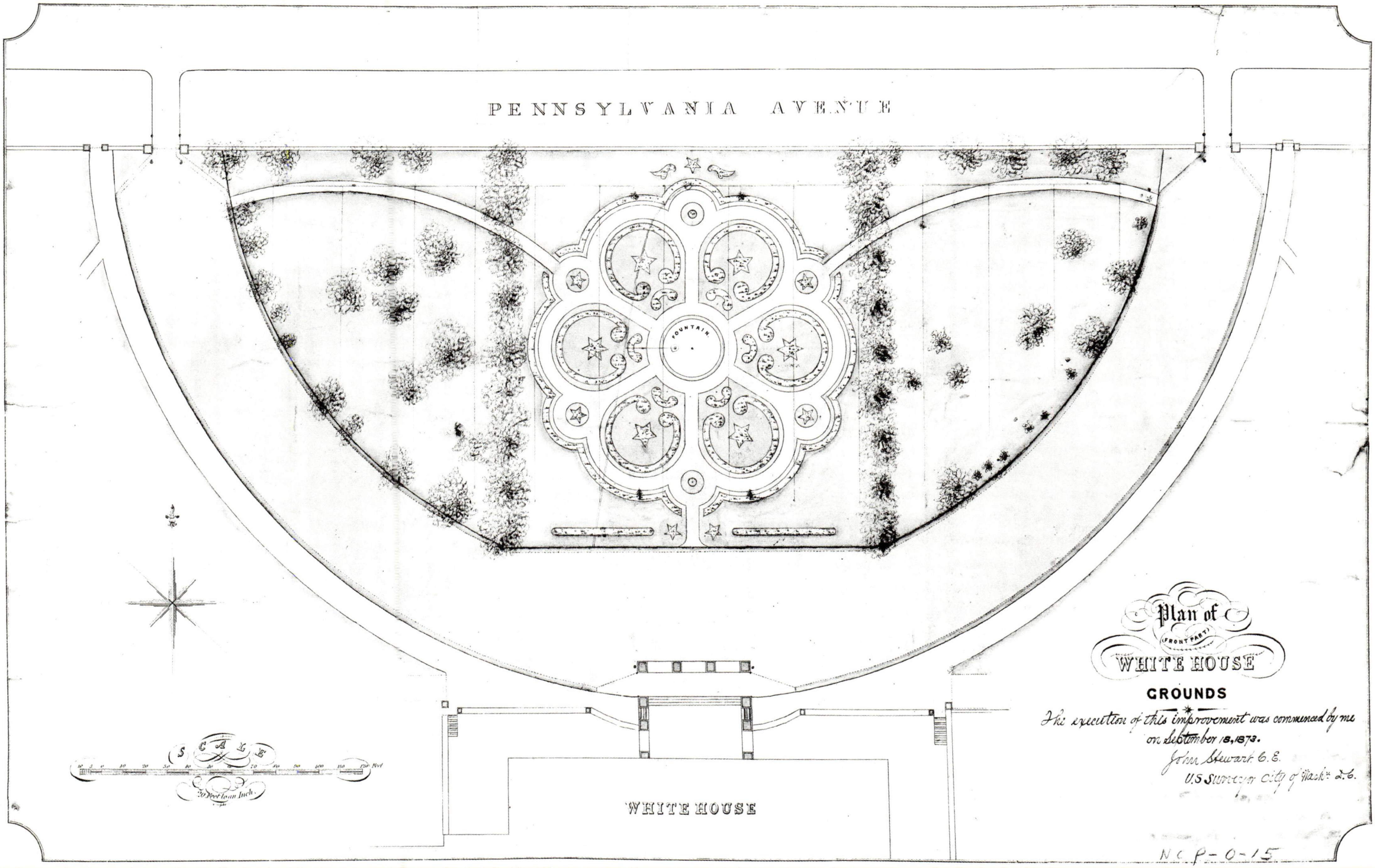
PENNSYLVANIA AVENUE
FOUNTAIN
Plan of
(FRONT PART)
WHITE HOUSE
GROUNDS
The execution of this improvement was commenced by me on September 18, 1873.
John Stewart C.E.
U.S. Surveyor City of Wash.n D.C.
SCALE
30 Feet to an Inch.
WHITE HOUSE
NCP-0-15

As a compromise between appearance and economy, the presidential gardeners began the tradition of setting tubs of live plants borrowed from the White House Conservatory and greenhouses around the fountain during spring, summer, and early fall. This yearly ritual continued for decades, with supplemental annual plantings also added at times according to the tastes of subsequent presidents and first ladies. These photographs from 1890 (above left) and 1897 (above right) capture the North Fountain surrounded by potted plants.

In 1873, just before the initial installation of the fountain, U.S. Surveyor John Stewart developed an elaborate planting arrangement that was a radical departure from the simple, green open space that had characterized the North Grounds for nearly three-quarters of a century. Stewart's grand planting scheme was in keeping with the lavish Victorian garden designs favored at the time, but it lasted only four years. Too costly to maintain, it was removed by 1877, the first year of Rutherford B. Hayes's presidency. But the president and his wife, Lucy Webb Hayes, shared a mutual interest in plants and gardening, and resolved that the North Grounds still needed flowers. As a compromise between appearance and economy, the presidential gardeners began setting tubs of live plants, borrowed from the White House Conservatory and greenhouses, around the fountain during spring, summer, and early fall. This yearly ritual continued for decades, with supplemental annual plantings added to accommodate the tastes of subsequent presidents and first ladies.

A special display was planted in the 1890s, when First Lady Frances Folsom Cleveland indulged her passion for floriculture by adding water lilies and other aquatic plants to all of the White House fountains, including the North Fountain. With the help of Head Gardener Henry Pfister, who had begun work at the White House at the beginning of the Hayes administration, she selected several species of lilies, including the giant water lily, which often grows to 6 feet in diameter and is capable of supporting a weight of 200 pounds. Mrs. Cleveland could easily have stood upon one of these immense floating pads, but history does not record if she was bold enough to do so.

President Theodore Roosevelt's Beaux-Arts–style remodeling of the White House in 1902 led to a more restrained approach to plantings around the North Fountain. Since the mid-twentieth century, the overall arrangement has been further streamlined and today typically includes only one or two types of annuals at a time. They are planted en masse and provide a vibrant burst of color. Seasonal replanting and careful plant selection help ensure that the beds are almost always in bloom, even in winter.

In 1902 President Theodore Roosevelt remodeled the White House in the restrained Beaux-Arts style, including the area around the North Fountain. In this B. M. Clinedinst photograph, gardeners can be seen planting seasonal flowers around the fountain (detail below). The equestrian statue of Andrew Jackson is seen across Pennsylvania Avenue in Lafayette Park.*

THE SOUTH GROUNDS

The President's Backyard

JUST AS THE NORTH GROUNDS serve as a public face or front yard to the White House, the South Grounds function as its private backyard. This long, gently sloping park-like expanse provides the first family with outdoor privacy and spaces for official government ceremonies and special events. As early as the first decade of the nineteenth century, Thomas Jefferson and his architect, Benjamin Henry Latrobe, envisioned the South Grounds as a "pleasure ground" reserved for the use of the president, his family, and invited guests. Jefferson's high stone wall provided a potent and easily understood means of restricting access, as did the massive triple-arched gate he had constructed at the far southeast corner of the site.

This drawing of the White House from the southwest by John Rubens Smith, c. 1833, shows the proportions of President Jefferson's high stone wall surrounding the South Grounds.

Presidents House Washington
103 11x16

A View of the White House *from the sketchbook of George Heriot, was made in 1815 in watercolor, with brown and gray ink and graphite, on paper. The artist captured visitors on the South Drive and the arch built by Thomas Jefferson at the southeast entrance to the grounds.*

Although the South Grounds were meant to be closed off, by the 1850s increasing numbers of people began filtering in from the publicly accessible North Grounds. The situation was manageable at first. But in 1873, President Grant began closing the entire grounds at sundown, a move criticized in some newspaper editorials, most likely influenced by Grant's political adversaries.

First Lady Julia Grant would have taken a stronger stance than her husband, as she valued her privacy and was bothered by crowds of "idle, curious, loungers" that gathered each afternoon on the South Grounds to watch her three children at play. Nellie Grant was fond of rolling her hoop, and Jesse loved to pedal his velocipede, while Buck, the oldest of the Grant children, practiced with his rifle. Had the same scene unfolded virtually anywhere else, it would have been mundane and forgettable. But this was the White House, and people came out to watch in droves. If Mrs. Grant had had her way, the president would have closed off the grounds morning, noon, and night. The fact that he did not suggests he was acutely aware that he was a tenant at the President's House, not the owner. Moreover, the possibility of being labeled as the man who shut the people out of the people's house was always a supremely dangerous political risk, and so Grant instead chose to try to balance his family's need for privacy with the public's desire to access the whole of the White House landscape.

Grant's successors followed his example, but some fared better than others in living in the spotlight. President Benjamin Harrison and his family seemed particularly unfazed by the attention in the early 1890s. At the time, there had not been young children at the White House for several years, and the president's namesake grandson, two-year-old Benjamin Harrison McKee, became something of a celebrity. "Baby McKee," as he was affectionately known, could oftentimes be found posing for photographs and shaking hands on the South Grounds, and the people loved him. At the other end of the spectrum, President Grover Cleveland ordered the South Grounds closed during his second term in order to keep his two young daughters "far out of reach of the ubiquitous Kodak" cameras. Public sentiment quickly turned against Cleveland, and not just in Washington. All across the country, people who had never been to the White House, and probably never would, felt evicted. A vicious rumor even began to swirl

President Benjamin Harrison's grandson, affectionately known as "Baby McKee," was one of the most photographed children at the White House. Baby McKee and his sister Mary were photographed here running on one of the South Lawn paths in 1889 as their nurse monitored their frolic.

alleging that "Baby Ruth," the Clevelands' eldest daughter, had been born with a deformity that the president wanted to keep secret. These charges were put to rest after Ruth appeared in public at the 1893 Easter Egg Roll.

Theodore Roosevelt tried a compromise. During the first two months of his presidency, he opened the South Grounds daily for a half hour in the morning and early evening, and also during lunchtime, with the stipulation that anyone availing themselves of this privilege would respect the household's right to privacy and refrain from snapping any photographs of the president or his family. Before long, however, this policy failed; people not only took photographs but expected the Roosevelt children to pose for them. Fed up, in November 1901 the president closed the South Grounds for the remainder of his tenure at the White House.

The matter was not definitively settled until 1923, when President Calvin Coolidge permanently closed off not only the South Grounds but the North Grounds as well. Security concerns had finally forced the issue, but privacy outdoors at the White House nonetheless remained hard to come by. Although people could no longer come and go as they pleased, there was nothing to stop curious onlookers or would-be photographers from peering into the president's backyard from outside of the fence.

Landscape architect Frederick Law Olmsted Jr. finally addressed this long-standing privacy problem in his 1935 report on the White House Grounds for President Franklin D. Roosevelt. In an effort to screen the area from view, a privet hedge had been planted around the perimeter of the South Grounds in 1909–1910 during the presidency of William Howard Taft. However, the hedge was always clipped just low enough so that people could see over it, and by the 1930s it was full of holes and scraggly. Olmsted filled in the gaps, allowed the hedge to grow, and accomplished the "double purpose of presenting a dignified and beautiful aspect to the public outside and at the same time securing effective privacy and seclusion within." He also went a step further and used the groves of fine, old trees growing on the South Grounds to create additional shady nooks and sheltered spaces removed from the public view. Olmsted's solutions were simple but extremely effective. He restored privacy to the president's backyard, and today this area remains a welcome oasis for the first family and their guests.

The Mystery of the Mounds

ABOUT HALFWAY DOWN THE SOUTH LAWN a pair of large knolls rises from the grassy green lawn. Their topography is unmistakable, and the mounds, as these large balls of earth are most commonly known, rank among the most iconic features anywhere within the presidential landscape. Given the great depth of study typically afforded to even the most ephemeral aspects of the White House, it comes as a considerable surprise that so little was known about the mounds for so long. The earliest definitive reference appears in an 1850 Washington guidebook that describes them as "artificial hillocks." This brief notation gave historians a date to work backwards from and an indication that the mounds are man-made constructions. The latter point is obvious enough. The mounds are too perfectly rounded, too evenly spaced, and too symmetrical to be natural occurrences. But who built them, and when? These answers proved frustratingly difficult to find.

An early view of the western mound was documented by Lefevre Cranstone in 1860 with a watercolor (left) entitled President's House, *while a 2012 photograph of the South Lawn captured the eastern mound (opposite).*

The National Archives maintains a vast collection of records—books, letters, receipts, bills, maps, drawings, photographs—dating all the way back to the groundbreaking for the White House in the 1790s, but these sources shed little light on the subject of the mounds. In the absence of hard evidence, the accepted theory credited Thomas Jefferson with their creation, and not without merit. While president, Jefferson initiated a large-scale earthmoving campaign on the White House Grounds as a necessary first step for future landscaping, and he seems to have favored a varied topography over a completely level lawn. One letter from Latrobe to Jefferson actually mentioned "a few gentle knolls & groves or clumps" on the South Grounds—a potential reference to the two mounds.

This theory held for many years, and eventually came to be taken as fact. Guidebooks and other publications on the White House began referring to the "Jefferson Mounds" on the South Grounds, but every now and then a new piece of information emerged that seemed contradictory. In the late twentieth century researchers uncovered correspondence dating to the 1850s that

The "Jefferson Mounds," identified by the colorized areas on this 1992 topographical map of the White House Grounds, were created between 1818 and 1824 during the Monroe administration.

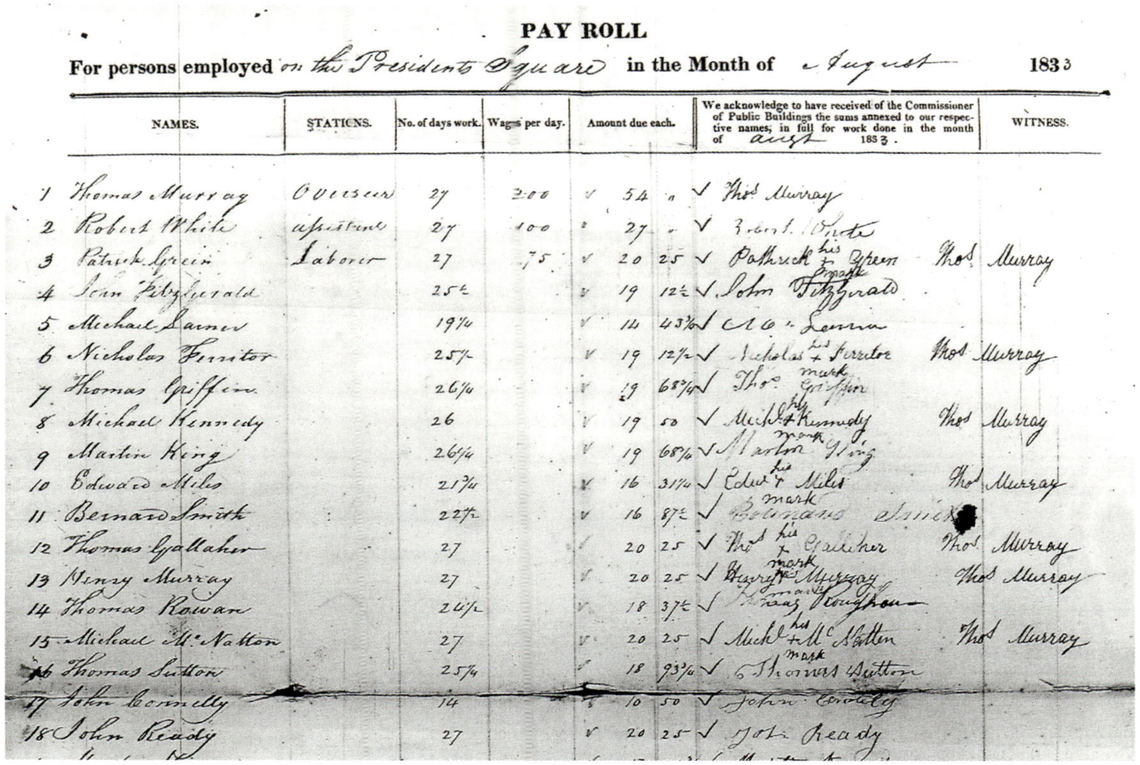
PAY ROLL

For persons employed on the Presidents Square in the Month of August 1833

NAMES.	STATIONS.	No. of days work.	Wages per day.	Amount due each.	We acknowledge to have received of the Commissioner of Public Buildings the sums annexed to our respective names, in full for work done in the month of August 1833.	WITNESS.
1 Thomas Murray	Overseer	27	2.00	54 "	Thos. Murray	
2 Robert White	assistant	27	1.00	27 "	Robert White	
3 Patrick Green	Laborer	27	.75	20 25	Patrick his x mark Green	Thos. Murray
4 John Fitzgerald		25½		19 12½	John Fitzgerald	
5 Michael Larner		19¾		14 43¾	[illegible]	
6 Nicholas Forrester		25½		19 12½	Nicholas his x mark Forrester	Thos. Murray
7 Thomas Griffin		26¼		19 68¾	Thos. Griffin	
8 Michael Kennedy		26		19 50	Michl. Kennedy	Thos. Murray
9 Martin King		26¼		19 68¾	Martin King	
10 Edward Miles		21¾		16 31¼	Edward his x mark Miles	Thos. Murray
11 Bernard Smith		22½		16 87½	Bernard Smith	
12 Thomas Gallaher		27		20 25	Thos. his x mark Gallaher	Thos. Murray
13 Henry Murray		27		20 25	Henry Murray	Thos. Murray
14 Thomas Rowan		24½		18 37½	Thomas Rowan	
15 Michael McNatton		27		20 25	Michl. his x mark McNatton	Thos. Murray
16 Thomas Sutton		25¼		18 93¾	Thomas Sutton	
17 John Connelly		14		10 50	John Connelly	
18 John Ready		27		20 25	John Ready	

The National Archives holds the payroll records for the 1818–34 earthmoving campaign, part of which was the creation of the mounds. The example at left is the "Pay Roll For Persons Employed on the Presidents Square in the Month of August 1833." It is interesting to note that many men simply signed their names with an "X," which was customary for those who were illiterate and unable to sign their own names. Beside these "X"s are signatures of witnesses and the notation "his mark."

referenced the planned creation of earthen mounds somewhere south of the White House. The "Pierce Mounds" theory might have supplanted the popular "Jefferson Mounds" theory, except that the 1850 Washington guidebook was published three years before Franklin Pierce came to the White House.

The eventual answer emerged from an unlikely combination of sources. In 1991, a large elm growing near the top of the eastern mound died and was cut down. When its growth rings were counted, the tree proved to be 165 years old and thus planted in 1825, during the presidency of John Quincy Adams. The mounds—or at the least the east mound—would have been present by this time, and supported the Jefferson theory nicely. An archaeological analysis of the mounds conducted in 1995, however, concluded that they were created c. 1818–24, during the Monroe administration.

Surviving historic documents also seem to support this date range. Voluminous payroll sheets show that a large workforce of manual laborers was engaged in the arduous task of grading the White House Grounds at this time. This work may have been a continuation of the earthmoving project begun under Jefferson, or it may have been required to remedy damage to the site incurred during the reconstruction of the White House after the British burned it in 1814. Regardless, the men followed a landscape plan, now lost, created by Architect of the Capitol Charles Bulfinch in 1818. Other than his predecessor Latrobe, Bulfinch was probably the most accomplished architect in the United States. In an era before landscape architecture was a profession, he was one of the few qualified to lead such an endeavor on the President's Grounds.

Because the Bulfinch plan is missing, there is no certainty. But the confluence of the tree ring data, archaeology, and historical documentation leaves no reason to doubt that the mounds were constructed between 1818 and 1824. They are therefore not the Jefferson Mounds. If anything they are the "Monroe Mounds," as he was president at this time, but the "Bulfinch Mounds" might be more precise.

Reshaping the Grounds: Expansion South

FOR MORE THAN HALF A CENTURY, the total extent of the White House Grounds was defined by the placement of Jefferson's wall. Roughly oval in shape, it enclosed some 15 acres of land and separated the grounds from the low-lying, marshy area that stretched farther south toward Tiber Creek and the Potomac River. Even after much of this wall was demolished and replaced by wooden perimeter fences in the mid-nineteenth century, the size and shape of the White House Grounds remained the same and followed the general outline of Jefferson's enclosure.

Almost immediately after the Civil War, however, things began to change. In the waning days of 1865, President Andrew Johnson granted Alfred B. Mullett, the government architect in charge of the extension of the U.S. Treasury Building next door, permission to proceed with a plan for laying down a new roadway. The future East Executive Avenue would run north to south between the Treasury and the White House. Mullett believed that the Treasury's expansive west wing, completed the year before, demanded such a thoroughfare so that the government clerks and visitors could fully appreciate its "grand and imposing appearance." The project was under way by early 1866, as workmen cut through the 50-foot-wide roadway as a southerly extension of Fifteen and One-Half Street (today's Madison Place). The following year, Mullett began planning another roadway as an analog to East Executive Avenue. West Executive Avenue would form a continuation of Sixteen and One-Half Street (today's Jackson Place) and separate the White House Grounds from the site of the future State, War and Navy Building (today the Eisenhower Executive Office Building), which Mullett was then designing. The work of establishing a proper grade and constructing West Executive Avenue did not begin until sometime in 1870 and continued for the next two years.

West Executive Avenue met East Executive Avenue in an arced half-circle containing about 5 acres of land south of the original boundary of the White House Grounds. The two new roads had significantly reduced the acreage of the presidential landscape on its east and west sides, and so it was decided to add this new southern enclosure as compensation. By June 1872, when the Executive Avenue project was essentially complete, laborers had already graded this southern addition so as to join the new South Grounds with the old. They also used cartloads of earthen fill obtained from the excavation of West Executive Avenue to change the original steep grade into a gentle slope down to the level of the Ellipse. The current shape and configuration of the White House Grounds were now in place and ranged over approximately 18.7 acres, up from the 15 acres or so that Jefferson knew. Moreover, the space now fit better within the overall tableau of President's Park, and thereby became the grand centerpiece of this larger landscape.

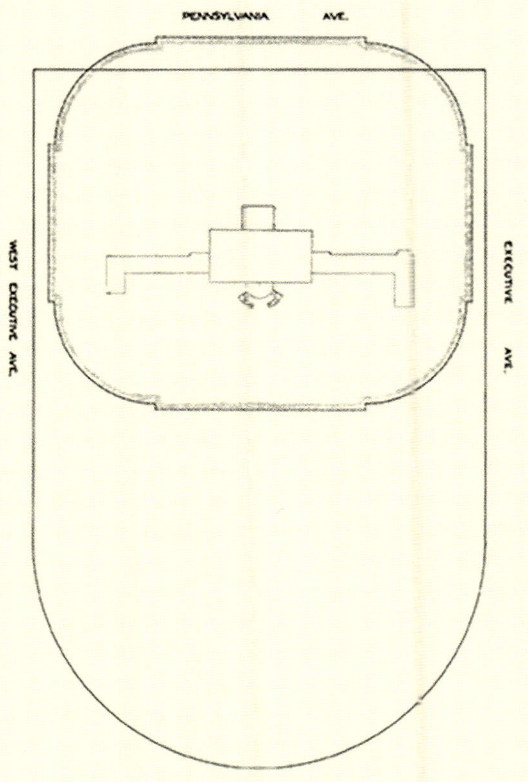

A National Park Service plan from 1927 shows the location of the original wooden fence built during the Jefferson administration in relation to the modern White House Grounds. The West Wing is not included in the drawing, as only the central block of the White House existed in 1801.

This landscape plan of President's Park, showing the extension of the South Grounds of the White House, is attributed to Alfred Mullett, c. 1870. Many elements of this drawing, including the walks on the extended South Grounds and Ellipse, and the building at the center of Lafayette Park, were never realized.

Winding Roadways

THE SOUTH GROUNDS encompass the largest portion of the entire White House enclave, and many landscape features and physical improvements have been constructed there. The hardscape of the South Grounds has varied extensively over time. Although some of its historic elements have come and gone, many roadways, fences, and ornamental water fountains remain in place today.

The bottom edge of this 1830 drawing of the South Grounds includes a hint of the east-west drive installed by Jefferson. This drive served as a public thoroughfare until the 1870s. The circular drive installed during the John Quincy Adams administration in 1825 can also be seen. It was covered with gravel for more than a decade before Jackson transformed it into a system of roads and enlarged it.

Jefferson envisioned roadways winding through mature trees and shrubbery in his c. 1807 plan for the White House Grounds, but he lacked the time and funding to carry out the project. Not seen in the drawing, the east-west drive would be beyond the southern boundary wall at the bottom edge of the plan.

Thomas Jefferson's correspondence and his c. 1807 landscape plan indicate that he intended to have an array of serpentine walks and roads installed on the South Grounds—a "pleasure garden"—but it is unlikely that the work was ever carried out. Jefferson did, however, have a road installed immediately south of the southern portion of his stone wall of enclosure. This road ran laterally east to west and served as a public thoroughfare. It was located beyond the boundary of the White House Grounds until the 1870s expansion, at which point workmen buried it under several feet of earthen fill during a regrading that also removed Jefferson's ha-ha. By 1872, they had installed a new east-west roadway approximately 15 feet to the north, where the new slope of the South Grounds could better accommodate it. With minor modifications, this transverse road is still in place today and marks the approximate southern boundary of the original White House Grounds.

A small circular driveway appeared on the near South Grounds in about 1825, during John Quincy Adams's presidency. Little is known about this drive except that it was covered with gravel and lasted for less than a decade before

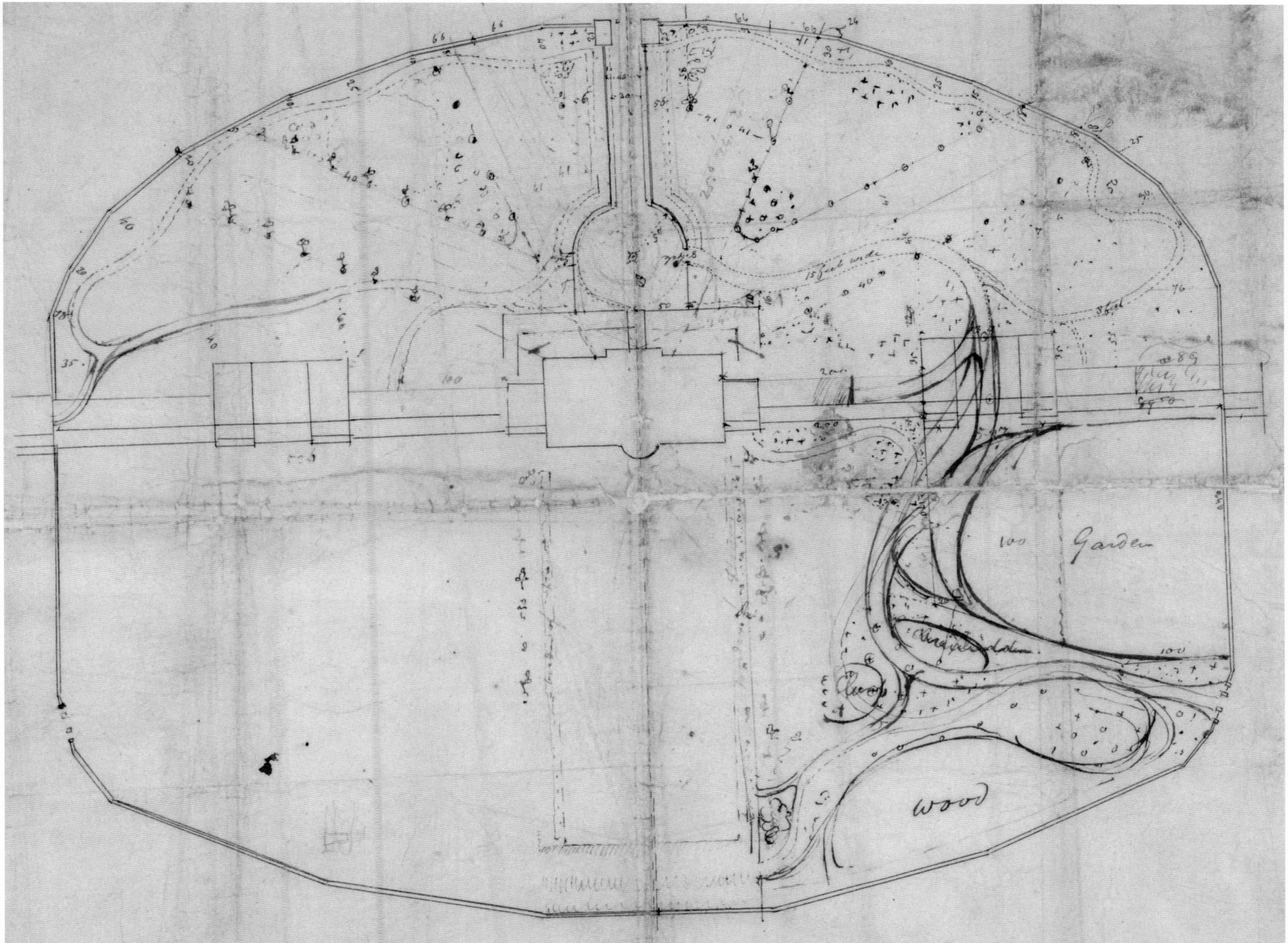
Garden
wood

ISOMETRICAL VIEW
OF THE
PRESIDENTS HOUSE
THE SURROUNDING
PUBLIC BUILDINGS
AND
PRIVATE RESIDENCES

This isometrical view of the White House from c. 1845–50 shows the distinctive guitar-shaped drive created by Andrew Jackson, as well as the mounds, and the kitchen and ornamental gardens. Jefferson's ha-ha wall and the east-west running road he constructed just outside the original southern boundary of the President's Grounds are visible at the far left of the image.

being replaced by a much more expansive system of roads and walks under Adams's successor, Andrew Jackson. Begun in 1833, this arrangement spread across most of the South Grounds and had a distinctive guitar-shaped roadway as its centerpiece, the curves of which arced around the famous pair of earthen mounds. Jackson also had a guitar-shaped driveway installed on the grounds of The Hermitage, his Tennessee home, upon his retirement in 1837. Legend holds that this drive imitated the lines of the prized guitar often played by the president's daughter-in-law Sarah Yorke Jackson, but both it and its counterpart on the White House Grounds may have been copying a similar driveway laid down on George Washington's Mount Vernon estate several decades earlier. Regardless, portrait painter Ralph E. W. Earl is usually credited with laying out the guitar-shaped drive at The Hermitage. Jackson's close friend, Earl lived at the White House throughout the president's two terms and would have had ample opportunity to redesign the South Grounds as well.

Downing's 1851 sketch plan for the White House landscape shows that he intended to replace the guitar and its associated roads and walks with a simpler eye-shaped configuration. But with his death, this change was not implemented, and Jackson's guitar remained at the center of the South Grounds for more than a century. It was finally removed in 1937 upon the advice of landscape architect Frederick Law Olmsted Jr.

Olmsted considered the guitar-shaped drive "a blemish on the landscape" and a "conspicuous interruption of the lawn." Despite his predilection toward preserving historic elements of the grounds, he urged President Franklin D. Roosevelt to have it "obliterated" and replaced by a simpler two-road circulation system that originated at the southeast and southwest corners of the grounds and gently curved up to meet one another before the South Portico. Olmsted felt that the transverse road constructed in 1872 ought to be removed as well, as it also caused an undesirable break on the lawn. President Roosevelt chose to implement Olmsted's recommendations for replacing the guitar-shaped drive but kept the old transverse road, as it provided a convenient shortcut across the South Grounds. Olmsted acquiesced, but sunk the roadbed down into the ground so that it became "invisible from the main floor of the White House." Overall, the system of roads on the South Grounds took on the shape of a circle with a flattened bottom, and two small triangles at the far southeast and southwest. This circulation system remains today and has been modified only slightly since the 1930s.

A series of sketches showing the evolution of the White House South Grounds, from 1851 to 1870, illustrates how Jackson's guitar-shaped drive remained at the center as the grounds evolved during the nineteenth century. It was finally removed in 1937 upon the advice of landscape architect Frederick Law Olmsted Jr. Note that not all the roads included in the 1870 sketch were actually constructed.

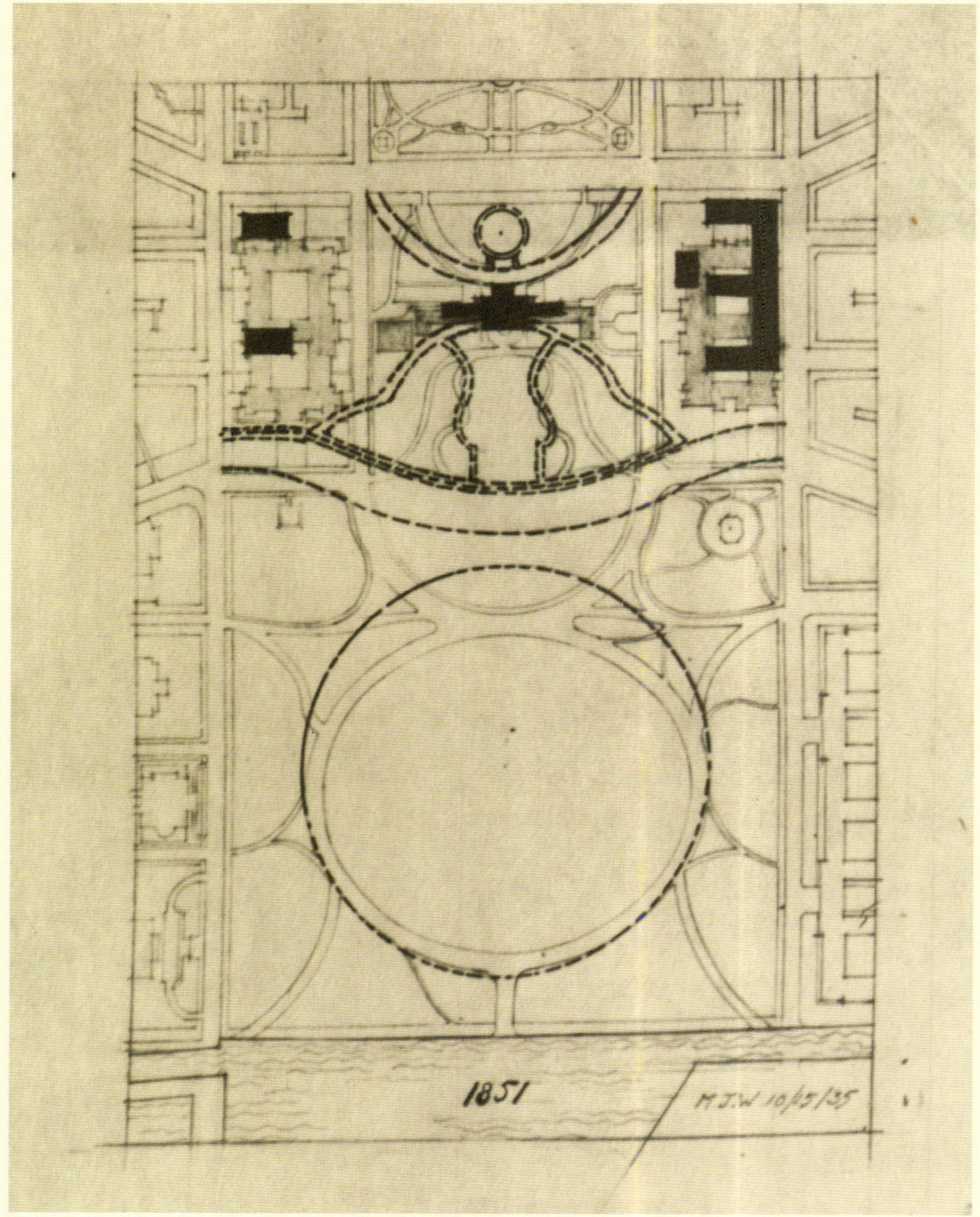

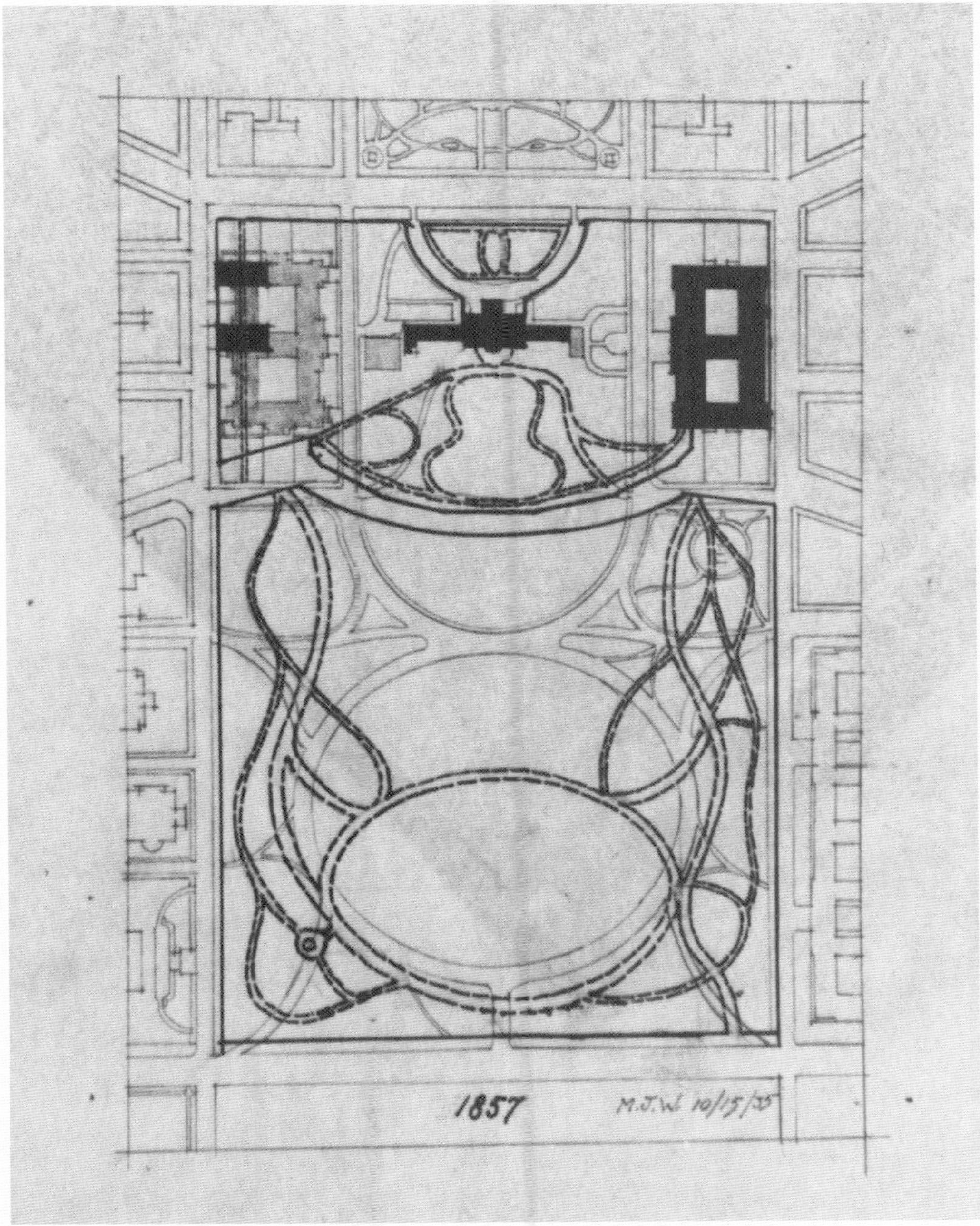
1857
M.J.W. 10/15/35

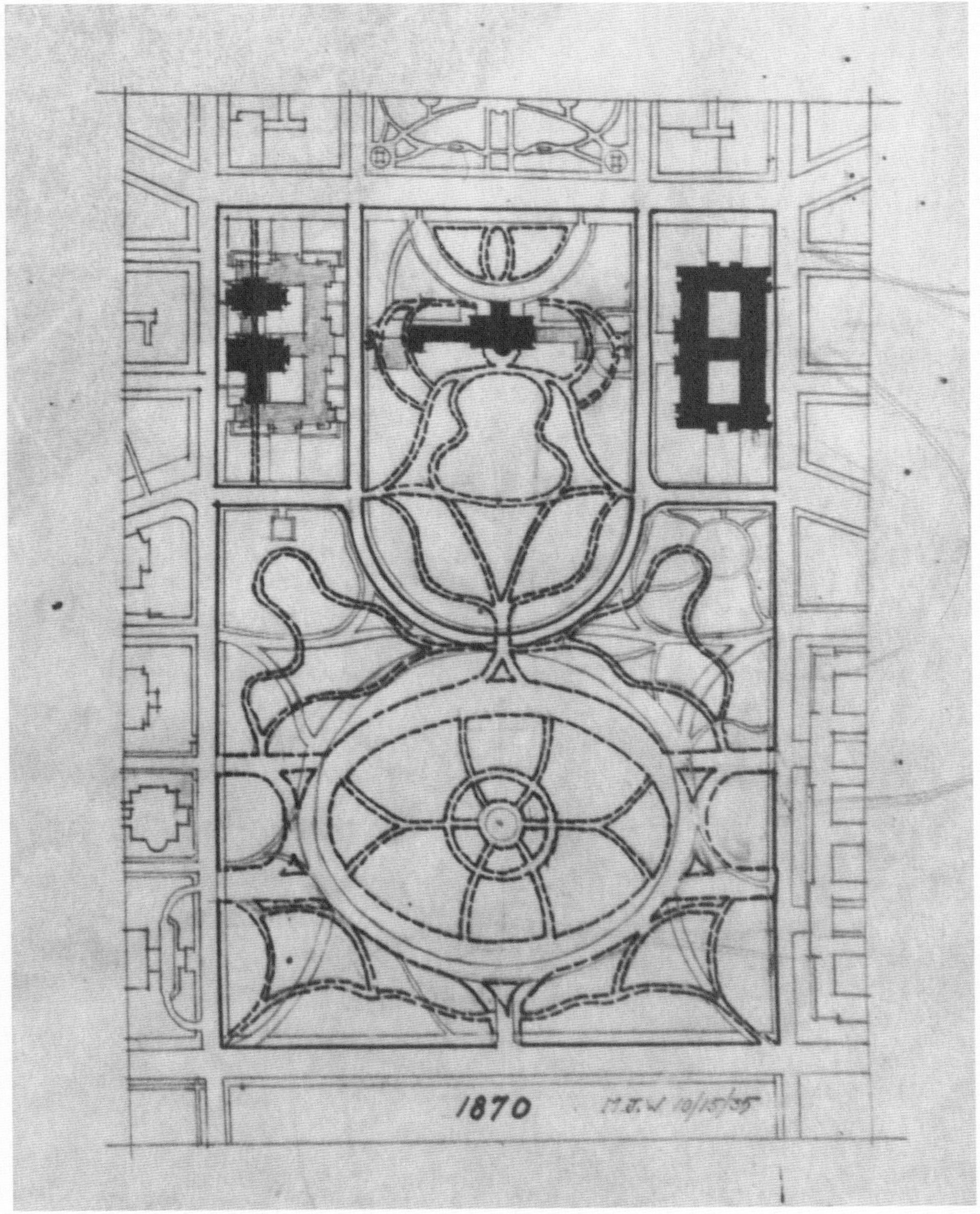
1870
M.J.W. 10/15/35

Fences

GOVERNMENT DOCUMENTS INDICATE that as early as 1828, major sections of Jefferson's stone wall were "going to ruin." As money was not forthcoming for repairs, much of the wall was torn down and replaced with runs of simple post-and-rail fences, which also needed to be repaired and replaced from time to time, resulting in a hodgepodge of materials enclosing the South Grounds by the time of the Civil War. Given the state-of-the-art security measures around the White House today, it is amazing to discover just how ineffectual these early barriers were. Cows grazed on the South Lawn so that first families could have fresh milk, but they sometimes wandered away by squeezing through a hole in the fence or pushing through a dilapidated gate. In 1857 a displeased secretary of war wrote a letter complaining that President James Buchanan's cows had escaped from the grounds and were wandering around outside his office. He requested, politely but firmly, that his boss corral these trespassers.

This drawing from c. 1828–33 shows the picket and railing fence from the southeast side of the White House.

This nineteenth-century engraving of the White House South Lawn depicts the simple post-and-rail fence that was installed after portions of Jefferson's stone wall were removed beginning in 1828. An ornamental garden is shown from the southwest instead of the southeast, where it was actually located on the grounds.

As part of the East Executive Avenue project, Alfred Mullett designed a low, neat wrought-iron picket fence that was installed atop a granite curb along the White House's side of the road by about 1866. When West Executive Avenue was cut through to meet East Executive Avenue in 1872, this style of fence was repeated along the entire perimeter of the expanded South Grounds. This short ornamental fence did little to enhance the security of the White House landscape, but it remained in place until Olmsted's revisions were made to the grounds in the 1930s, when a taller iron fence modeled after Paulus Hedl's original design along Pennsylvania Avenue took its place. This change gave a uniform appearance to the perimeter fencing enclosing the White House Grounds, with the older portion to the north dating to the 1820s and 1830s and the newer sections around the South Grounds to 1937–38.

Mullett's low iron fence and gate are seen in this photograph of the South Facade of the White House, c. 1872. This fence extended along the perimeter of the South Grounds and was more of an ornamental feature than a security barrier, as the fence is used today.

In March 1962, a passerby was able to give a treat to Caroline Kennedy's pony Macaroni through the fence surrounding the South Grounds.

Today, security considerations outweigh the aesthetics of White House fencing. The post-9/11 barriers that supplement the fencing around the South Grounds include concrete, wire, and chains.

Fountains

THE FIRST ORNAMENTAL WATER FOUNTAIN in the White House landscape was installed on the near South Grounds by 1858, more than a decade before the larger North Fountain. Built as part of a project to update and expand the plumbing at the house by drawing water from the Potomac River, the whimsical showpiece incorporated a circular basin and two terra-cotta bowls supported by a group of stylized creatures that have alternately been described as dolphins and sea serpents. Unfortunately, the plumbing was faulty, and this malfunctioning fountain was rarely used. Under President Ulysses S. Grant, a simple rock pile design replaced the sea serpents in 1869, and a new water jet was added in 1875, but it remained notoriously temperamental. President Rutherford B. Hayes removed it in 1878 to make room for a croquet court.

A detailed rendering dating from the time of the Civil War depicts the sea serpent fountain on the South Lawn when it was new to the landscape.

The newly installed water fountain on the South Grounds, comprised of two tiers and a shaft of entwined sea serpents rising over a large round pool, can be seen here in this stereograph from c. 1867. It was built ten years before the North Fountain when the plumbing at the house was expanded, but it malfunctioned often and was rarely used.

Children during Lincoln's administration pose around the sea serpent fountain in this carte de visite, c. 1861–65. In 1869 the sea serpents were replaced with a simple rock pile design; a water jet was added in 1875.

N.Y. Av.
G. St.
F. St.
Pa. Av.
E. St.
D. St.
O. Av.
C. St.
B. St. North
13½ St.
15½ St.
16½ St.
17th St.
Pa. Av.
G. St.
F. St.
N.Y. Av.
E. St.
D. St.
C. St.
B. St.

The grand circular basin installed on the lower South Grounds in 1876 can be seen in this 1877 plan. It measured 75 feet in diameter and 3½ feet deep (nearly twice the size of the North Fountain, which also appears in the plan) and was the second largest fountain in the city.

Water lilies planted by Head Gardner Henry Pfister are seen growing in the South Fountain in this photograph, c. 1890. The rim of the bowl lay level with the ground, making the entire fountain appear to float on a sea of green grass. A 2½-foot border of white cobblestones set in cement made an accommodating bench for those willing to risk getting wet, and an "eel trap" installed in the underground supply pipe prevented unwanted fish, rocks, and rubbish from getting into the fountain along with the water being piped in from the Potomac.

The South Grounds were not without a fountain, however, as a grand circular basin had been added to the southern extension in 1876. Originally fitted with the French gurg coupling, the South Fountain received a new nozzle after the gurg was removed and reinstalled in the North Fountain. In 1886 the South Fountain gained an additional sixteen water jets. Around this time Head Gardener Henry Pfister fitted the basin with eight brick pockets for growing water lilies and other aquatic plants, which First Lady Frances Cleveland made full use of during her years at the White House in the 1880s and 1890s.

The South Fountain remains largely unchanged since Mrs. Cleveland's lilies bloomed there, and is today an iconic feature of the South Grounds. Every president since Grant has appreciated it, but two short, amusing anecdotes are particularly worth repeating. For a time in the late 1800s the fountain was stocked with goldfish and perhaps other types of fish as well. According to a period newspaper article, on at least one occasion President Chester Arthur went fishing here, and "had good luck." The article may amount to nothing more than a presidential fish story, but the account is at least plausible as Arthur was in fact an accomplished fisherman. Forty years later, President Warren Harding frequently teed off near the South Portico and drove golf balls the length of the grounds into the fountain below.

The South Vista

No discussion of the south grounds of the White House would be complete without mention of the splendid view down from the South Portico. Termed the South Vista, in Jefferson's time this sightline overlooked nothing but marshy swamplands reaching down to the swift waters of the Tiber Creek and the Potomac River, yet he saw the potential for something grander. In order to keep this view unobstructed, Jefferson had the southernmost portion of his high stone wall built as a sunken ditch known as a ha-ha.

In this 1862 carte de visite by Mathew Brady, Union soldiers walk atop the stone ha-ha wall built by Jefferson to terminate the South Grounds. The ha-ha remained in place until 1871, when it was demolished.

The ha-ha wall can be seen running just below the center of this 2011 painting by Peter Waddell entitled Tiber Creek: The Bathers: John Quincy Adams Takes a Deadly Chance. *The painting tells the story of a near drowning in 1825 in the fast-moving Tiber when President John Quincy Adams took a chance by trying out an abandoned boat. The boat promptly began to sink, and the president was rescued and pulled to shore. Today the Tiber flows safely beneath a masonry vault, over which passes Constitution Avenue.*

This photograph taken from the roof of the White House in 1946 shows how clear the view was across the South Vista toward the newly built Jefferson Memorial after Olmsted ordered dense foliage to be cleared.

Ever since the South Vista was restored, every president has enjoyed the splendid view of the Washington Monument and Jefferson Memorial from the back door of the White House.

Downing shared Jefferson's affection for the South Vista. His 1851 landscape plan shows the view unbroken—with the exception of one seemingly out of place tree—through the South Grounds and out over his planned Parade Ground, which became under President Hayes the present-day Ellipse. Even farther south stood the unfinished nub of the Washington Monument, located at what was then the farthest extent of the National Mall. In essence, this is the same spatial relationship that exists today, and the South Vista provides the sole tangible link between the White House Grounds as Downing envisioned them and the present landscape, now more than a century and a half later.

However, during the second half of the nineteenth century into the first quarter of the twentieth, tree planting on the South Grounds gradually encroached upon the scene and obstructed this cardinal sightline. By the time Frederick Law Olmsted Jr. began his work on the White House Grounds, the view was almost entirely blocked. In response, Olmsted initiated a selective thinning of the foliage in order to reopen the historic South Vista, which he considered not only an integral component of the presidential landscape but also a scene of "marked aesthetic importance to the general plan of the National Capital" that ranked among Washington's most "distinctive and inspiring features."

Ever since its restoration, every president has marveled over the splendid view. The completion of the Jefferson Memorial in 1943 extended the axis of the South Vista to the domed peristyle on the far side of the Tidal Basin. Four years later, the bronze statue of President Jefferson was added to the memorial. Fittingly, he looks north toward the White House, back along the same sightline that he created so long ago.

The Children's Garden

PRESIDENT AND MRS. LYNDON B. JOHNSON bestowed what is perhaps the greatest gift given to the White House Grounds by any administration: the Children's Garden. Set off to the west side of the South Grounds so as not to interrupt the South Vista, the Children's Garden is a diminutive 18 by 30 feet and private. Its location ensures that it blends seamlessly with its surroundings and grants the delightful experience of being in a wooded park area, just as its designers, Edward Durrell Stone Jr. and Edward Alexander, intended. Overall, the space is defined by a Pennsylvania blue flagstone terrace set into the ground in an irregular pattern. Within this terrace is a fishpond measuring 4 by 8 feet across and 2 feet deep that is kept stocked with goldfish in warm weather. Child-size cast-iron chairs and a matching table show that the garden, as evidenced by its name, is meant to be enjoyed by children.

A plaque in the garden inscribed "To the White House Grounds from the President and Mrs. Lyndon B. Johnson, Christmas 1968" marks the entry to the Children's Garden that they established.

A winding flagstone path leads into the Children's Garden.

A Stayman winesap apple tree, easily climbable, is the only tree of any size. Other plantings include narcissus, grape hyacinths, crocuses, azaleas, mountain laurel, periwinkle, and Mondo grass. A plaque inscribed, "To the White House Grounds from President and Mrs. Lyndon B. Johnson, Christmas 1968," is imbedded in the flagstone walk leading to the little play area. The president and first lady formally dedicated the Children's Garden on January 19, 1969, Johnson's last full day in office. At the time, the bronze footprints of Lyndon Patrick Nugent and the handprints of Lucinda Desha Robb, the president's grandchildren, were already laid into the flagstone terrace. Since then, grandchildren of every subsequent president have added either their handprints or footprints, creating a touching visual record of the young family members that have come to stay at the White House since the mid-twentieth century.

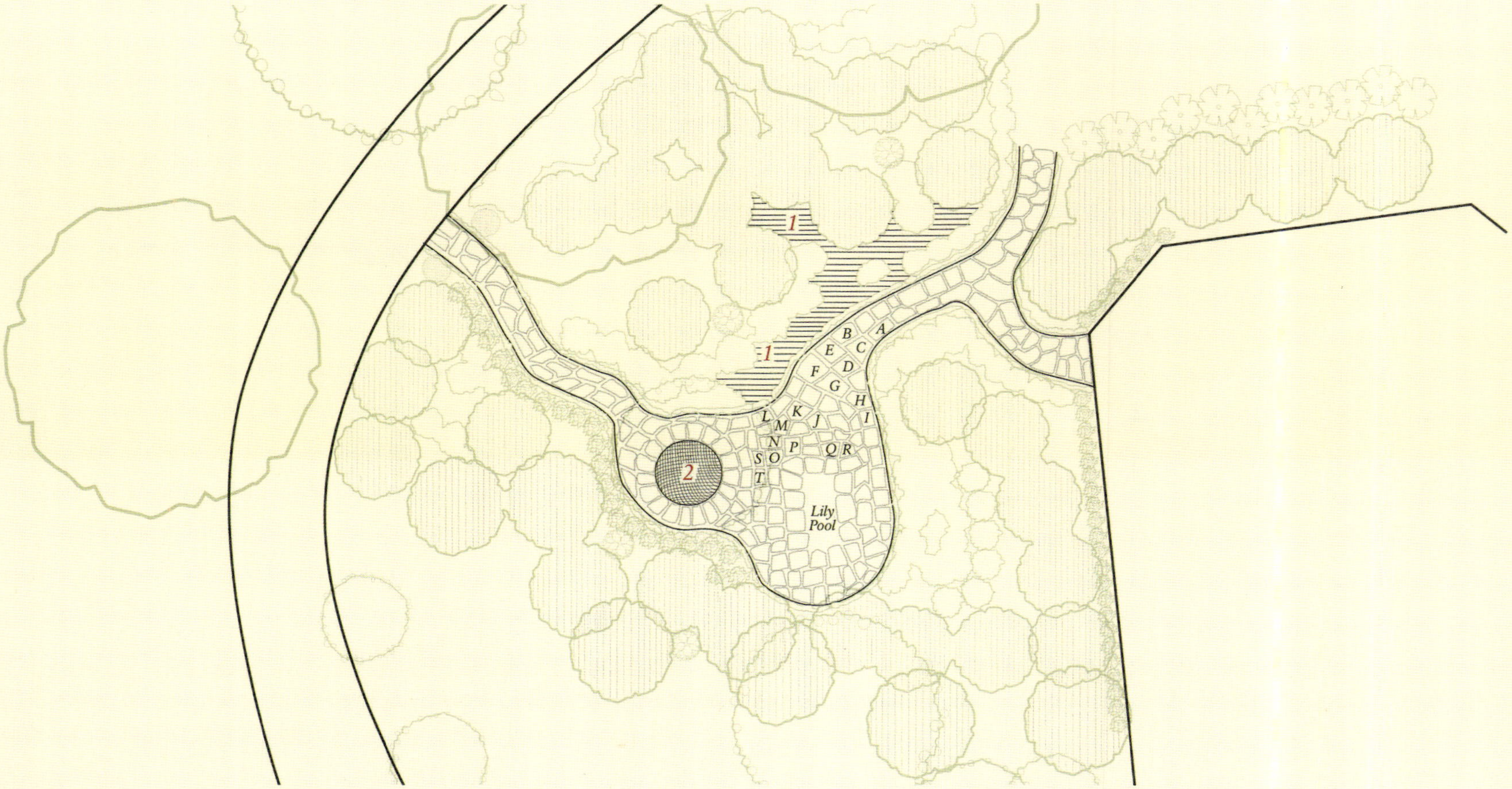

Plaques

A TO THE WHITE HOUSE GROUNDS FROM THE PRESIDENT AND MRS. LYNDON B. JOHNSON, CHRISTMAS 1968
B ELLIE LE BLOND
C SAM LE BLOND
D PATRICK LYNDON NUGENT
E LUCINDA DESHA ROBB
F JEBBY BUSH
G GEORGE P. BUSH
H NOELLE BUSH
I JAMES EARL CARTER IV
J SARAH ROSEMARY CARTER
K MARSHALL BUSH
L WALKER BUSH
M JASON JAMES CARTER
N ASHLEY BUSH
O PIERCE BUSH
P LAUREN BUSH
Q JENNA BUSH
R BARBARA BUSH
S CAMERON REAGAN
T ASHLEY MARIE REAGAN

Seasonal Plantings

1	NARCISSI	*Daffodil 'Naturalizing Mix'*	Combination of colors	Mid spring
2	BLUE PANSY	*Viola X wittruckiana 'Crown Blue'*	Deep blue	Mar. - Apr.

This map of the Children's Garden produced by the National Park Service in c. 2008 shows the location of spring plantings and the arrangement of the handprints and footprints made by presidential grandchildren.

A terrace surrounds a small fishpond in the center of the Children's Garden. During warm weather the fishpond is stocked with goldfish and lily pads.

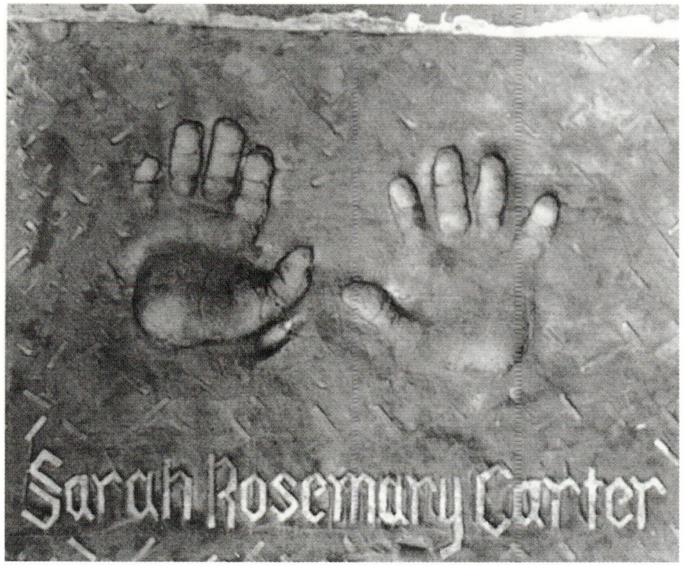

Since the first hand- and footprints were made by President Johnson's grandchildren, seventeen other presidential grandchildren have pressed their hands and feet into clay to be cast in bronze and set in the garden pathway. Among them are Sarah Rosemary Carter, President Jimmy Carter's granddaughter (handprints at left) and Lauren Bush Lauren, President George H. W. Bush's granddaughter (posing in the garden at right).

THE WEST GARDENS

PRIOR TO THE TURN OF THE TWENTIETH CENTURY, the areas flanking the South Portico were decorated with flowerbeds and other ornamental plants in keeping with Victorian gardening practices. While pretty, these plots were small and rather unimaginative, a far cry from the celebrated gardens that now grace the near South Grounds of the White House. The space on the west side was also occupied by an extensive greenhouse complex that supplied flowers for the interior rooms of the mansion as well as bulbs, annuals, and perennials for planting throughout the grounds. When the greenhouses were demolished in 1902 to make way for the new West Wing offices, most of the hothouse plants were lost, but the opportunity to establish a formal garden outside the president's office was created. First Lady Edith Roosevelt developed a colonial-style garden in the space, and a few years later First Lady Ellen Wilson modernized its design. In 1962, during the Kennedy administration, the space was redesigned and became the Rose Garden we know today.

The First Ladies' West Gardens

EDITH ROOSEVELT'S colonial garden bore little resemblance to the current Rose Garden. Intricate and heavily embellished, it was modeled after the geometrically patterned gardens arranged outside large estates such as Mount Vernon in the late eighteenth century. Symmetrical flower parterres, surrounded by close-clipped boxwood hedges and crossed by a web of graveled walks, formed the skeletons of both Mrs. Roosevelt's West Garden and a complementary garden she placed east of the South Portico. Both bloomed with old-fashioned flowers.

The West Garden parterres, seen in the photographs at left and below and in the diagram opposite, were laid out as two large, four-petaled flowers and contained roses, evergreen shrubs, clematis, ferns, palms, flowering herbs, and old-fashioned flowers such as snapdragons, mullein pinks, cardinal flowers, goldenrods, bugloss, boneset, columbines, phlox, milfoil, China aster, poppies, syringa, larkspur, ambrosia, hollyhocks, Jessamine, and peonies. Observers noted a preponderance of yellow, Edith Roosevelt's favorite color.

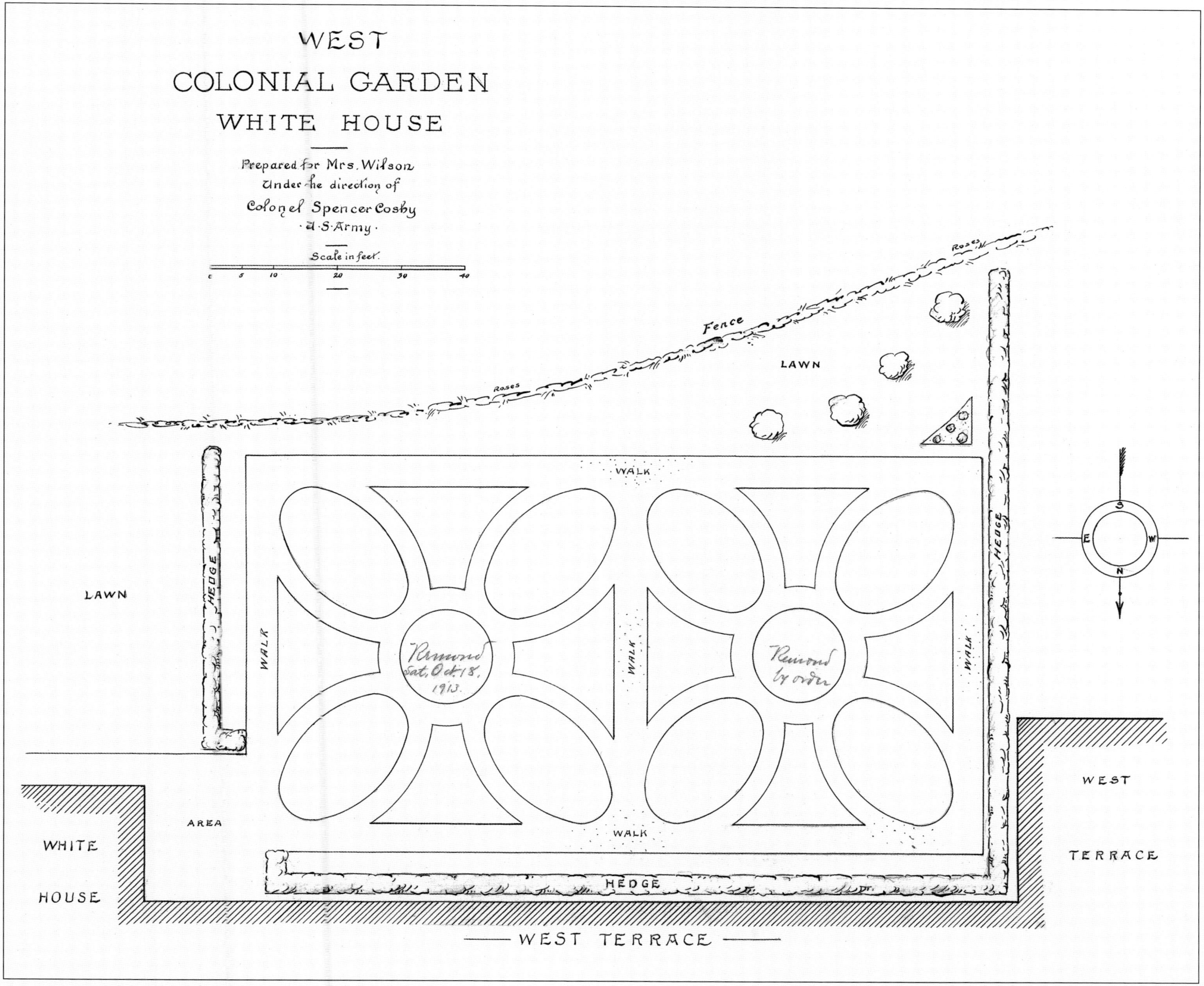
WEST
COLONIAL GARDEN
WHITE HOUSE
Prepared for Mrs. Wilson
Under the direction of
Colonel Spencer Cosby
U.S. Army.
Scale in feet.
0 5 10 20 30 40
Roses
Fence
Roses
LAWN
WALK
HEDGE
S
E
W
N
LAWN
HEDGE
WALK
Removed Sat. Oct. 18, 1913.
WALK
Removed by order
WALK
WEST
TERRACE
AREA
WHITE
HOUSE
WALK
HEDGE
WEST TERRACE

This drawing by George Burnap, a government landscape architect, shows his plans for executing Ellen Wilson's vision for the West Garden.

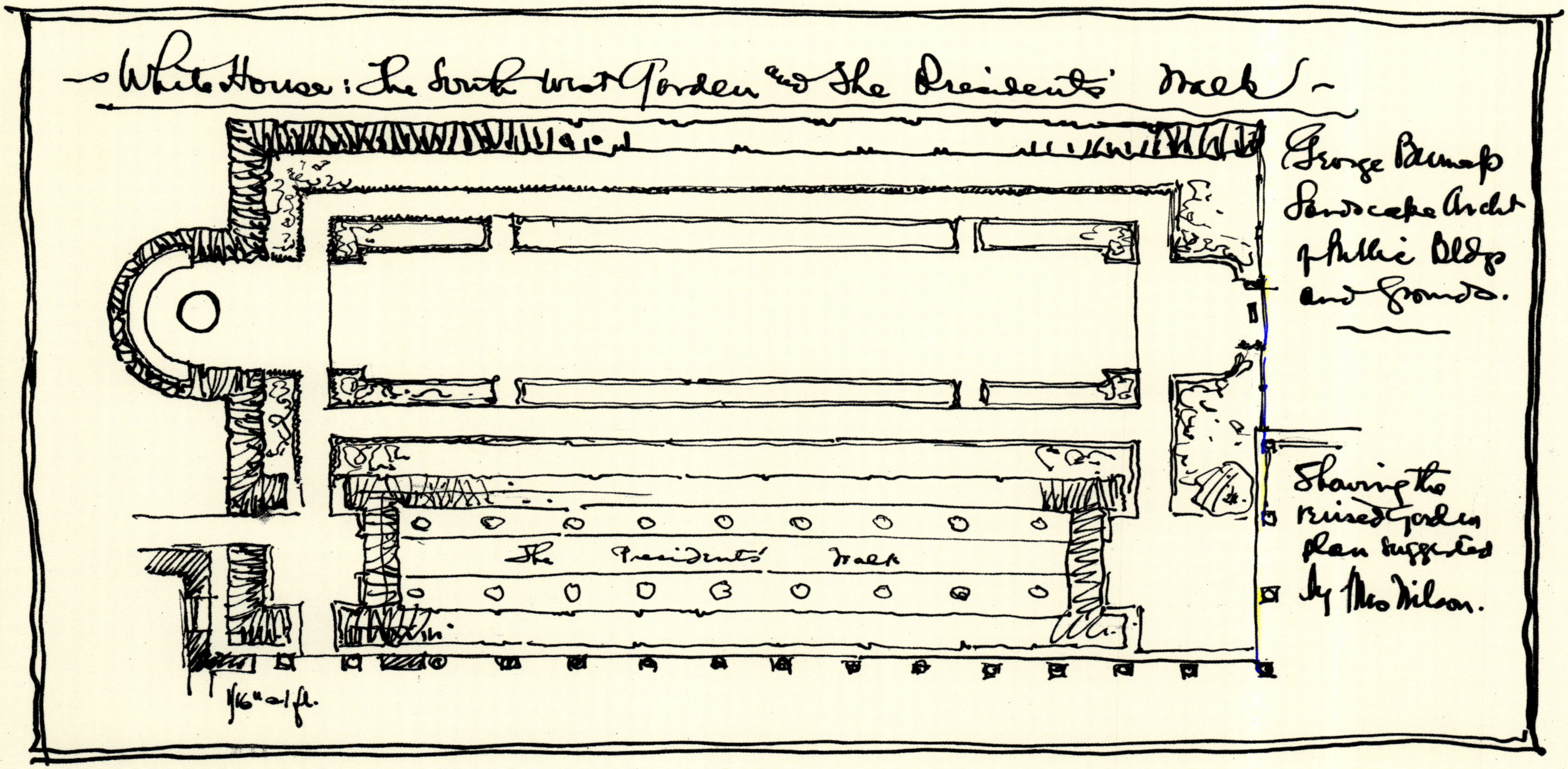

This 1927 aerial view of the South Grounds captures Ellen Wilson's mature gardens. On the left of the South Portico is the West Garden, designed by George Burnap, and on the right of the portico is the East Garden, designed by Beatrix Farrand.

Completed in the spring of 1904, the two colonial gardens survived for less than a decade. Looking out from her bedroom window one day in 1913, Ellen Axson Wilson, the first wife of President Woodrow Wilson, gazed down upon her predecessor's work and was not impressed. A skilled garden designer and artist whose watercolors often featured landscapes, Mrs. Wilson considered colonial gardens outdated and unimaginative. Rather than looking to the past for inspiration, she embraced the new formalism in garden design that stressed clean lines and clarity.

Mrs. Wilson chose George Burnap, a government landscape architect working in Washington, to execute her vision for the new West Garden. Burnap is largely unknown today but in the early twentieth century he was at the top of his profession. Within Washington, he is perhaps best known for adorning the Tidal Basin at First Lady Helen Taft's behest with the famous Japanese flowering cherry trees, but his projects include the designs of the East Potomac, Farragut, Garfield, Lincoln, Logan, Montrose, and Meridian Hill Parks; Mount Pleasant Triangle; Thomas and Washington Circles; and three gardens on the south lawn of the Treasury Building.

At the White House, Burnap's new West Garden appeared as a rectangle with sharp corners and long, unobstructed views across open grass panels and straight planting beds containing specimen rosebushes, narcissus, pansies, and a variety of spring bulbs. High boxwood hedges marked the garden's southern and eastern boundaries and functioned as privacy screens. Burnap also created a special walkway running between the presidential offices in the West Wing and the Executive Residence in the main block of the White House. Dubbed the "President's Walk," this tree-lined allée formed the northern edge of the garden and passed just south of the colonnaded West Terrace built in Jefferson's time and still in use as a service wing during Wilson's tenure at the White House.

Burnap's plan for the West Garden included a statue of the Greek god Pan and an elaborate latticework pergola at the far west end. He placed a white bench and table designed in the American craftsman style at the garden's opposite east side.

This photograph taken in 1913 shows Ellen Wilson's finished formal garden. The "president's walk" is seen on the far right.

When a terminally-ill Ellen Wilson inspected the Pan statue, it was her final venture beyond the White House walls.

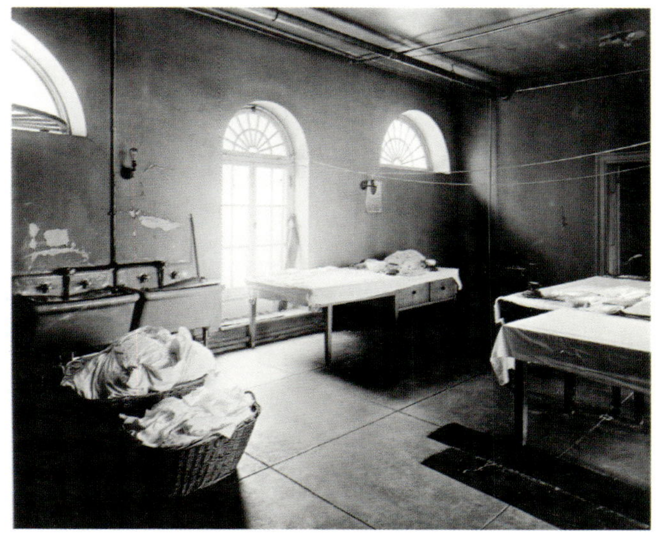

The idea for a special walkway connecting the West Wing to the Executive Residence occurred to Ellen Wilson as a way to have her husband spared the ignominy of having to walk back and forth through a pantry and past the laundry each day.

The Mid-Century West Garden

TRAGICALLY, ELLEN WILSON had very little time to enjoy her new outdoor space. Following a fall in her bedroom and a brief battle with Bright's disease, she died on August 6, 1914. The West Garden remained largely as she left it until the 1930s, when it began to show signs of decline. Some of the plants failed to thrive while others were overgrown. Despite its worsening condition, the garden apparently received little attention during World War II and suffered more damage during President Harry S. Truman's full-scale renovation of the White House.

This photograph of the West Garden taken in 1935 by Frederick Law Olmsted Jr. shows how Ellen Wilson's rose garden had declined. Deferred maintenance during the Great Depression led to an overall appearance best described as disheveled or makeshift.

At the high noon of the Truman renovation, the gardens have disappeared as the grounds have become a stage for the monumental construction project.

Between 1948 and 1952, the 150-year-old White House was gutted to its exterior walls for long overdue structural repair and stabilization. During this time the grounds—particularly on the south—became a vast construction site. Upon completion of the Truman renovation, the West Garden was replanted with a profusion of azaleas and more than 1,400 new rosebushes. The gardeners utilized popular floribunda rose varieties as well as a selection of smaller hybrid tea roses. Overall, the West Garden resembled Ellen Wilson's garden, only simpler, and with less variety in it's plantings.

Truman's successor, President Dwight D. Eisenhower, replaced some of these roses with yet more azaleas. They were pink, of course, the well-known favorite color of First Lady Mamie Eisenhower. Otherwise the overall character of the West Garden changed very little during Eisenhower's eight years in office.

Throughout the Truman renovation, a bewildering assortment of building materials, debris, and temporary construction sheds littered the South Grounds. The president's backyard was, for a time, virtually unrecognizable.

After the Truman renovation was finished in 1952, the same familiar boxwood hedges remained around the perimeter, but not so tall as before, supplemented here and there with replacements as needed.

Burnap's old garden seat remained in place for a time, but the Pan statue was long gone. Head Gardener Robert Redmond is seen here inspecting the new floribunda roses in May 1953, a little more than a year after they were planted.

The Kennedy Rose Garden

IN AUGUST 1961, President John F. Kennedy sat down with Rachel Lambert Mellon, a personal friend with a background in horticulture, and asked her to redesign the West Garden as a place of beauty and one that might welcome 250 guests. Mellon had already awoken within the president a previously unknown interest in gardening by giving him a copy of Thomas Jefferson's published garden diaries, and Kennedy now offered her the chance of a lifetime. After a brief moment of trepidation, she accepted the commission and quickly enlisted the help of two professionals: Perry Wheeler, a friend and landscape architect living and practicing in Washington, and Irvin Williams, a National Park Service horticulturist, hired in 1949, and based out of the Kenilworth Park and Aquatic Gardens in Northeast Washington, D.C. He had begun working on projects at the White House in 1956.

Prior to Rachel Lambert Mellon's redesign, long rows of privet divided the West Garden and limited the number of guests that could be accommodated at such events as President Truman's welcome to a group of journalists, pictured here in 1951.

President Kennedy envisioned a garden that would in a sense be a green theater, with open space to accommodate guests and wide steps that would serve as a presidential stage. Rachel Mellon's success in creating the theater President Kennedy envisioned is apparent in this recent peek into the garden (opposite, c. 2014), which reveals the timeless perfection of the concept.

At the request of Mellon and Wheeler, Williams was transferred to the White House full time to take over the care and management of the grounds. Over the years, he took on the titles of head gardener, horticulturist, and superintendent. His long and celebrated tenure continued for nearly half a century, a truly remarkable term of service that far exceeds that of any other caretaker and spanned every presidential administration from John F. Kennedy to George W. Bush. His expertise still casts a long shadow over the White House gardens.

Mrs. Mellon's selection of plants ensured that the garden was almost perpetually in bloom: bulbs in the spring; lavender, heliotrope, geraniums, lilies, and dianthus in the summer; daisies in the fall; and hollies, periwinkle, and the boxwoods providing winter greenery. A planting list dated June 15, 1962, when most of the plants had been installed, provides the fullest description of the floral richness of the garden during Kennedy's time at the White House. The following is a transcription of the original, with the numbers of individual plants by species given in parentheses: saucer

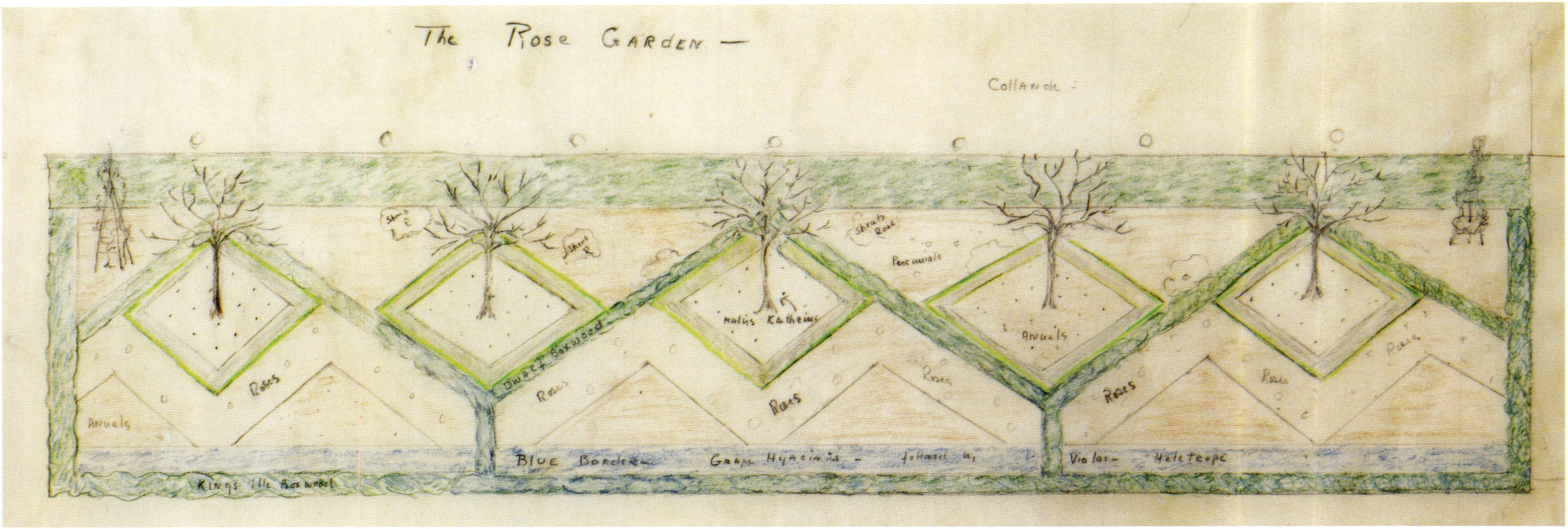

Rachel Mellon's sketch shows her plans for boxwood-bordered flowerbeds to be planted on either side of the central lawn. Katherine crabapple trees were to be planted at the center of each of the diamond parterres. The parterres themselves were defined by a thick planting of santolina, which in turn gave way to successive arrangements of blooming flowers and shrubs.

Mellon's 1962 watercolor shows her design concept for an open lawn of 50 feet by 100 feet flanked to the north and south by the 12-foot-wide boxwood-bordered flowerbeds.

magnolias (4); "Katherine" crabapples (10); Washington hawthorn (1); geraniums—"Enchantress" (310), "Pink Cloud" (310), "Red Irene" (305); rosebushes—hybrid tea "Speaker Sam" (9), hybrid tea "The Doctor" (9), hybrid tea "Frau Karl Druschki" (9), hybrid tea "Peace" (9), floribunda "World's Fair" (9); petunias—"Blue Magic" (92), "Purple Luster" (58), "Popcorn" (112); Dianthus—"Mrs. Sinkins" (231), "Bravo" (300); yarrows—Achillea taygetea (20), Achillea filipendulina "Gold Coronation" (20); heliotrope "Vanilla Peruvianum" (400); phlox "Mia Rugs" (20); Nicotiana (112); hosta "Virdis Marginata" (56); alpine barrenwort (250); Sedium sieboldii "Stonecrop" (300); gaillardia "Burgundy Blanket Flower" (72); primroses (52?); Shasta daisies (2); Santolina incana (36). The plants whose quantities were not specified are as follows: holly hedges (Osmanthus ilicifolius and Ilex acquifolium), English boxwood borders (Buxus sempervirens "Suffruticosa" and Buxus sempervirens "Green Pillow"), periwinkle vines, and Korean spice viburnum.

The arrangement was breathtaking, but as the name implies the roses were the stars of the Rose Garden and they produced a profusion of color. Though made famous by Rachel Mellon, roses were not new additions to the space. They had been present in Edith Roosevelt's colonial gardens and in the West Garden that Burnap designed for Ellen Wilson. Roses had also been grown elsewhere on the White House Grounds during the latter half of the nineteenth century, and Rachel Mellon chose to celebrate this history. As she explained, "The one flower that unites all the occupants through the history of the White House is the rose. Thus, for most of the twentieth century, the Rose Garden has been a rose garden. Now, in 1961, President Kennedy wanted it restored in spirit but revised to become more than just a private garden." The White House Rose Garden, as the world knows it today, fulfills his wish.

for The President's Garden
The White House
-1962-

An Outdoor Oval Office

ADJACENT TO THE WEST WING, and steps from the Oval Office, the Rose Garden is an extension of the president's workspace. Woodrow Wilson set up an outdoor office there during the summer of 1914, when his wife was terminally ill and World War I was erupting in Europe. He canceled a family vacation so the first lady could rest while he concentrated on the worsening international situation. As the stress took a toll on his own health, Wilson's doctors urged him to get more fresh air and relaxation. Unwilling to leave the White House, he struck upon the novel idea of moving the executive branch outdoors. The president ordered a large fly tent set up over part of the West Garden that had only recently been finished according to his wife's specifications. He added electric lights and telephone service and that summer operated almost exclusively out of his "open air office." Everything from cabinet meetings on the war to one-on-ones with senators and congressmen to his more mundane duties took place under the shade of the canopy.

President Wilson's outdoor office, shown from above. Note the fly tent over the bench and table designed by George Burnap.

President Wilson's tented "Outdoor Oval Office," c. 1914.

President Warren G. Harding frequently used the West Garden as a backdrop for official photographs, such as this one taken during a visit by members of Native American tribes in 1921.

By the Eisenhower administration, as seen in this photograph from 1959, the Wilson-era garden had been modified. The size of the flowerbeds was reduced, and lawn area was expanded by 30 percent, thereby opening up enough space for an additional sixty guests.

Wilson's successor, President Warren G. Harding, started two new traditions by using the West Garden as a floral backdrop for official photographs and as a ceremonial space to greet dignitaries and special guests. Subsequent chief executives followed Harding's example, and President Eisenhower particularly enjoyed using the space as a setting for speeches and state ceremonies. The central lawn within the garden served as a convenient place for the audience to assemble, either standing or seated on folding chairs, while the president spoke from the colonnaded West Terrace or from within the garden itself.

John Kennedy had little more than a year in which to enjoy his new Rose Garden, before his assassination on November 22, 1963. Nonetheless, he made full use of this brief time, employing the outdoor area as a venue for his addresses, bill signings, and other functions, as well as a private space where he could speak in confidence and without reserve to even the most high-profile visitor to the White House. Thus it was in the garden that he privately confided to the Reverend Dr. Martin Luther King Jr. that the Federal Bureau of Investigation had placed him under surveillance.

On April 9, 1963, President Kennedy addressed his guests at a Rose Garden ceremony to bestow honorary U.S. citizenship on Winston Churchill.

In June 1971, President Richard Nixon's daughter Tricia was married to Ed Cox in a Rose Garden ceremony attended by four hundred guests. The garden was specially decorated for the occasion with white tree roses, Madonna lilies, heliotrope, pink geraniums, and white petunias planted along its border.

The annual Thanksgiving presidential turkey pardoning is traditionally held in the Rose Garden. Seen here in 2010, President Obama pardoned a turkey named Apple with daughters Malia and Sasha by his side.

After acceding to the presidency in 1963, Lyndon Johnson continued to use the Rose Garden as a stage. There he signed a number of groundbreaking bills into law, including his War on Poverty program and other aspects of his larger Great Society initiative. Among these were several bills pertaining to conservation, city beautification, and natural resource stewardship, all fitting topics for the Rose Garden setting.

On June 12, 1971, President Richard Nixon's daughter Tricia married Edward Cox in the Rose Garden. The gala event marked the first marriage ever held on the White House Grounds and combined the pomp and circumstance associated with the most ceremonial state functions and the charm and joy of a typical American wedding. Capitata yews adorned the South Portico steps down which the bride and her father descended arm-in-arm. To satisfy the immense public interest in the event, the wedding was broadcast live across the United States, with camera crews filming the ceremony from a platform farther down the South Lawn.

During the late twentieth century, the Rose Garden was also used by presidents as a place for speaking directly to the American people, and this practice continues today. In recent memory are President George W. Bush's Rose Garden press conferences following the September 11, 2001, terrorist attacks and during the subsequent wars in Iraq and Afghanistan.

In contrast to serious news conferences is the annual "pardoning" of a Thanksgiving turkey, a lighthearted tradition that dates to the Lincoln administration. The event often takes place in the Rose Garden.

THE EAST GARDENS

THE ENCLOSED SPACE ON THE EAST SIDE of the South Portico has followed a similar evolution to that on the west, beginning with sequential gardens by First Ladies Edith Roosevelt and Ellen Wilson, and culminating in the 1960s with a major redesign by Rachel Lambert Mellon. Today the East Garden is called the Jacqueline Kennedy Garden, dedicated by First Lady Lady Bird Johnson on April 22, 1965, to her predecessor. Tucked between the South Portico and the East Wing, it is viewed and enjoyed very often by the president's family and is a beautiful tapestry of color for visitors to look out upon as they walk through the East Colonnade toward the main block of the White House.

The First Ladies' East Gardens

FIRST LADY EDITH ROOSEVELT's east colonial garden complemented its counterpart to the west and was composed of the same boxwood hedges and old-fashioned flowering plants. The only significant difference between the two gardens when planted in 1904 lay in the shape of the geometric parterres formed by the hedges and gravel walks. In the East Garden, they resembled four-pointed snowflakes and contained a number of arrow-shaped planting beds.

First Lady Edith Roosevelt posed for portrait painter Theobald Chartran with the South Grounds as a backdrop in 1902.

Edith Roosevelt's colonial East Garden as it appeared shortly after its completion in 1904.

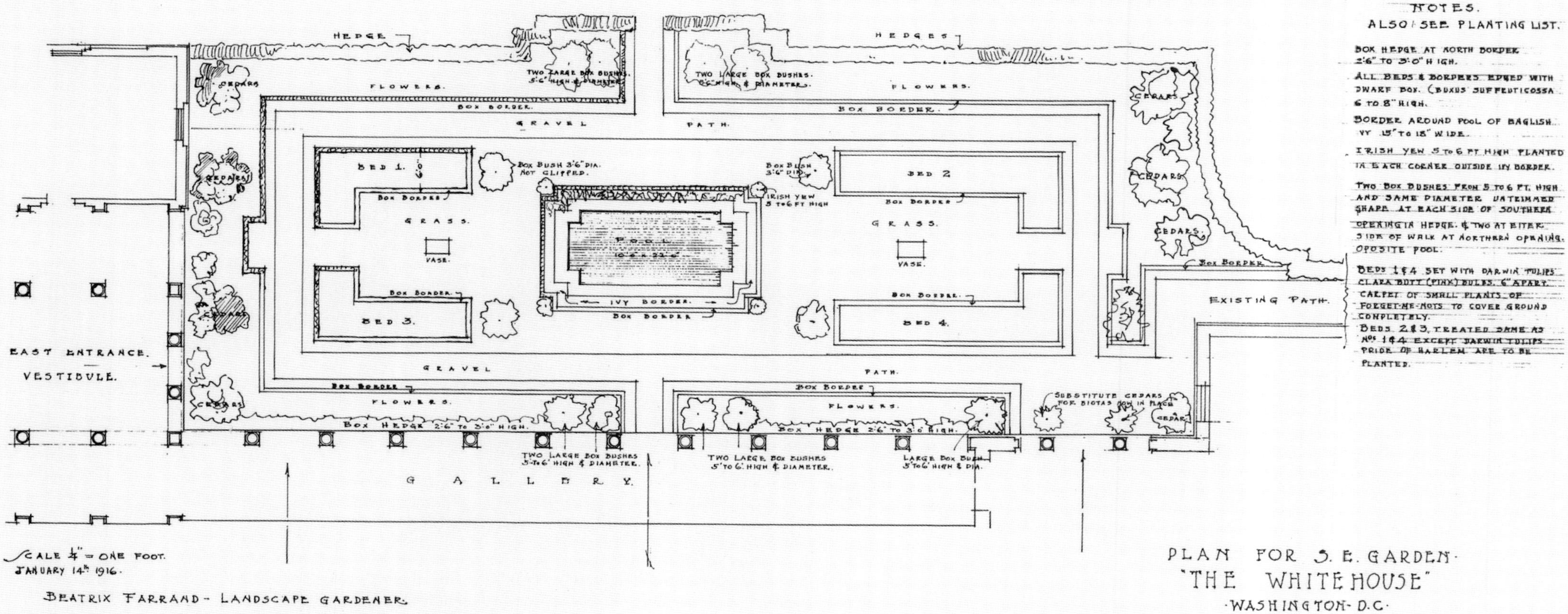

Within the East Garden, two cultivated varieties of Darwin tulips, "Clara Butt" and "Pride of Helen," dominated the flowerbeds atop a lush groundcover of forget-me-nots. There were also larkspurs, pinks, candytufts, Sweet William, and lavender bushes that together created a soft pastel color palette. Beatrix Farrand placed an Irish yew tree at each of the pool's four corners and mixed several other small evergreen trees in with the boxwoods along the garden's perimeter, as seen in this planting plan from 1916.

The East Garden designed by Beatrix Farrand for First Lady Ellen Axson Wilson in 1913 was not installed until after Mrs. Wilson's death. It was President Wilson's second wife, Edith Wilson, who oversaw its installation in 1916. The mature garden is seen here in the early 1920s. The image on the right is a hand-colored photograph taken by Frances Benjamin Johnston in the spring of 1921.

For her redesign of the East Garden, First Lady Ellen Wilson selected landscape architect Beatrix Jones Farrand, one of the most talented professionals of her generation, a charter member of the American Society of Landscape Architects, and the only woman among the society's eleven founders. She is best known for her design of the Dumbarton Oaks estate in Georgetown, the Reef Point Gardens in Bar Harbor, Maine, the Henry Huntington Museum in Pasadena, California, and several American college and university campuses. Most of these commissions, however, came after her design for the White House, and it is likely that Farrand's professional standing improved as a consequence. Although she had been practicing for almost twenty years, she had formerly been better known for her poignant and informative writings on the subject of garden design than for the projects she had actually undertaken. Farrand cultivated her unique aesthetic by blending a myriad of interests and sources of inspiration, including fine art. She was, in essence, an artist who worked in plants, water, and rocks. The landscape was her medium, and she was therefore an ideal choice for the artistically inclined Ellen Wilson.

Farrand developed her design for the East Garden concurrently with George Burnap's submissions for the West Garden. There is no evidence that they collaborated, but the overall coherence of their two gardens indicates that they were at least aware each other's work. Farrand's design was straightforward and uncluttered and also featured the same clean lines and crisp right angles found in the West Garden. The familiar boxwood hedges and planting beds were here as well. Evergreens were substituted for Virginia cedars, due to their lower cost. The only wholly new element introduced by Farrand was also the East Garden's principal feature: the large, rectangular lily pool situated prominently at its center. A suite of garden furniture designed by Farrand including chairs, loveseats, stools, and tables was placed in the garden, each piece featuring a naturalistic pattern of grapevines, tree branches, and cornstalks as well as animals and birds.

The Mid-Century East Garden

FARRAND'S EAST GARDEN was completely destroyed during the Truman renovation, its site used to hold construction equipment and temporary buildings. After the renovation was completed in 1952, National Park Service gardeners put down a layer of fresh topsoil and replanted the garden with 1,000 sweet olive evergreen shrubs, 1,000 boxwoods, 300 azaleas, thousands of Dutch tulip bulbs, and an unspecified number of hyacinths and other tulips. Speed was of the essence, as Queen Juliana of the Netherlands would arrive on April 4, making an official state visit to dedicate the Netherlands Carillon just across the Potomac River in Arlington, Virginia. Afterward, a reception was planned for her in the East Garden, the first major function held on the grounds in nearly three years.

A view of the South Front of the White House during the Truman renovation captures the East Garden completely cleared of trees and plants to allow for construction equipment and supplies.

This view of the South Front of the White House, taken April 4, 1952, shows the East Garden quickly planted with flowers in bloom to welcome Queen Juliana of the Netherlands.

To meet this deadline, all the flowers and flowering shrubs were planted in-bloom, and the Dutch tulips, which were selected in homage to the queen and her people, were forced from bulb to flower in a mere twenty-one days by the staff of the National Capital Parks. To save time, the flats in which they had been forced were imbedded in the soil to give the impression that the tulips had been planted in the garden the previous fall. Thanks to this trick and hard work, the garden was ready to greet the visiting monarch.

The Jacqueline Kennedy Garden

After becoming first lady in 1963, Lady Bird Johnson initiated a complete renovation of the East Garden, which remained much as Truman had left it. She called on Rachel Mellon, fresh from her success with the Rose Garden, for this project as well. Although far from an exact copy of the Rose Garden, Mrs. Mellon's finished creation had much in common with her previous work. The new East Garden, soon christened the Jacqueline Kennedy Garden, was completed in the fall of 1964 and featured lateral planting beds arranged to the north and south of an open, central lawn. A row of European linden trees was planted along the East Terrace of the White House, and topiary hollies set off a wide assortment of seasonal flowers. A white grape arbor designed by the now world-renowned architect I. M. Pei ran along the garden's west end, and a small fountain and sculpture of a girl holding a plant and watering can by the artist Sylvia Shaw Judson punctuated its east side.

In December 1965, Rachel Mellon received the Department of the Interior's Conservation Service Award in recognition of her designs for the Rose Garden and Jacqueline Kennedy Garden. Shortly after their move into the White House in 1981, President Ronald Reagan and First Lady Nancy Reagan asked Mrs. Mellon, then in her 70s, to refresh the Jacqueline Kennedy Garden and Rose Garden, which she did without undertaking any significant changes to her original designs.

Mrs. Johnson's interest in redesigning the East Garden related more broadly to her conservation and beautification efforts, particularly her leadership of the Society for a More Beautiful Capital, which planted millions of flowers in the public spaces of Washington, D.C. She became the first first lady to actively advocate for legislation by supporting the Highway Beautification Act of 1965. "Lady Bird's Bill," as it was known at the time, encouraged the scenic enhancement of the Interstate Highway System and other important national roads, as well as stricter limitations on billboards placed along them.

The East Garden was dedicated as the Jacqueline Kennedy Garden on April 22, 1965. In this photograph of the event, guests gather in front of the new grape arbor designed by architect I. M. Pei. First Lady Lady Bird Johnson (in yellow) is seen beneath the new arbor.

First Lady Hillary Clinton poses in the Jacqueline Kennedy Garden behind Untitled *by artist Joel Shapiro in a 1994 installation. A June 1998 installation (right) included the sculpture* Bird Effigy *by Truman Lowe (in the foreground),* Guardians & Sentinels *by Susie Berins Ericsen/ Qimmiqsak (right rear), and* Earth Messenger Totem *by Doug Coffin (standing in the background).*

During her time at the White House, Mrs. Johnson often held planning meetings as well as teas and other small parties in the Jacqueline Kennedy Garden, and most of her successors have availed themselves of the space for similar purposes. The bones of the garden have not changed significantly in the half century since Rachel Mellon implemented her design. However, the botanical composition of the plants growing there has varied tremendously according to the tastes of every first lady. Even more than the Rose Garden, which remains focused primarily on its namesake roses, the East Garden functions as a place of self-expression for the women (and men) living in the White House. The annual spring and fall garden tours, held each year since the Nixon administration, provide an ideal opportunity for the public to see any new additions in bloom.

Perhaps no one utilized this space to a greater extent than First Lady Hillary Rodham Clinton. In the spring of 1994, Mrs. Clinton, as honorary chair of the Committee for the Preservation of the White House, initiated a program to display contemporary American sculpture in the Jacqueline Kennedy Garden. The idea came to her shortly after moving into the White House when she discovered, much to her surprise, that the art collection contained few sculptural pieces. She later wrote, "That seemed a notable oversight, and so I became committed to bringing sculpture to the White House for our millions of visitors and guests."

An exhibition opened in October 1994 with the first twelve sculptures. Every six months to a year thereafter, a dozen new pieces, loaned from the finest museums and collections around the nation, were installed on a rotating basis. The focus was on contemporary art after the 1930s. The exhibits were spearheaded by Committee for the Preservation of the White House members Kaki Hockersmith and J. Carter Brown, director of the National Gallery of Art (1969–92), in consultation with the Association of Art Museum Directors and with the generous support of the Iris and B. Gerald Cantor Foundation. Among the many sculptors represented were Deborah Butterfield, Alexander Calder, Willem de Kooning, Roy Lichtenstein, Paul Manship, Louise Nevelson, Isamu Noguchi, Georgia O'Keeffe, George Rickey, and David Smith. More than 10 million visitors enjoyed the White House Sculpture Garden before it ended with the Clinton administration in 2001.

First Lady Michelle Obama welcomes guests to a White House reception in the East Garden to celebrate the reopening of the White House Visitor Center in September 2014. The grape arbor designed for the East Garden by I. M. Pei in 1965 remains in place as seen below in a recent springtime photograph.

VISITING THE WHITE HOUSE

The East Entrance

MOST OF THE 10 MILLION people who came to see First Lady Hillary Clinton's White House Sculpture Garden in the 1990s passed through the mansion's East Entrance, which has served as the primary visitor gateway into the White House and grounds since the close of the Civil War. In 1865 it was clear that, after years of hard use and lack of upkeep, Jefferson's old East Terrace was crumbling and too far gone to save. By mid-June 1866 President Andrew Johnson had had the terrace razed, the rubble hauled away, and remaining foundations covered over with topsoil and graded. The east side of the main block of the White House then came back into view, and it appeared quite diminutive compared with the west facade of the hulking Treasury Building. Even worse, there was now a large empty space between the two buildings, creating an imbalance with

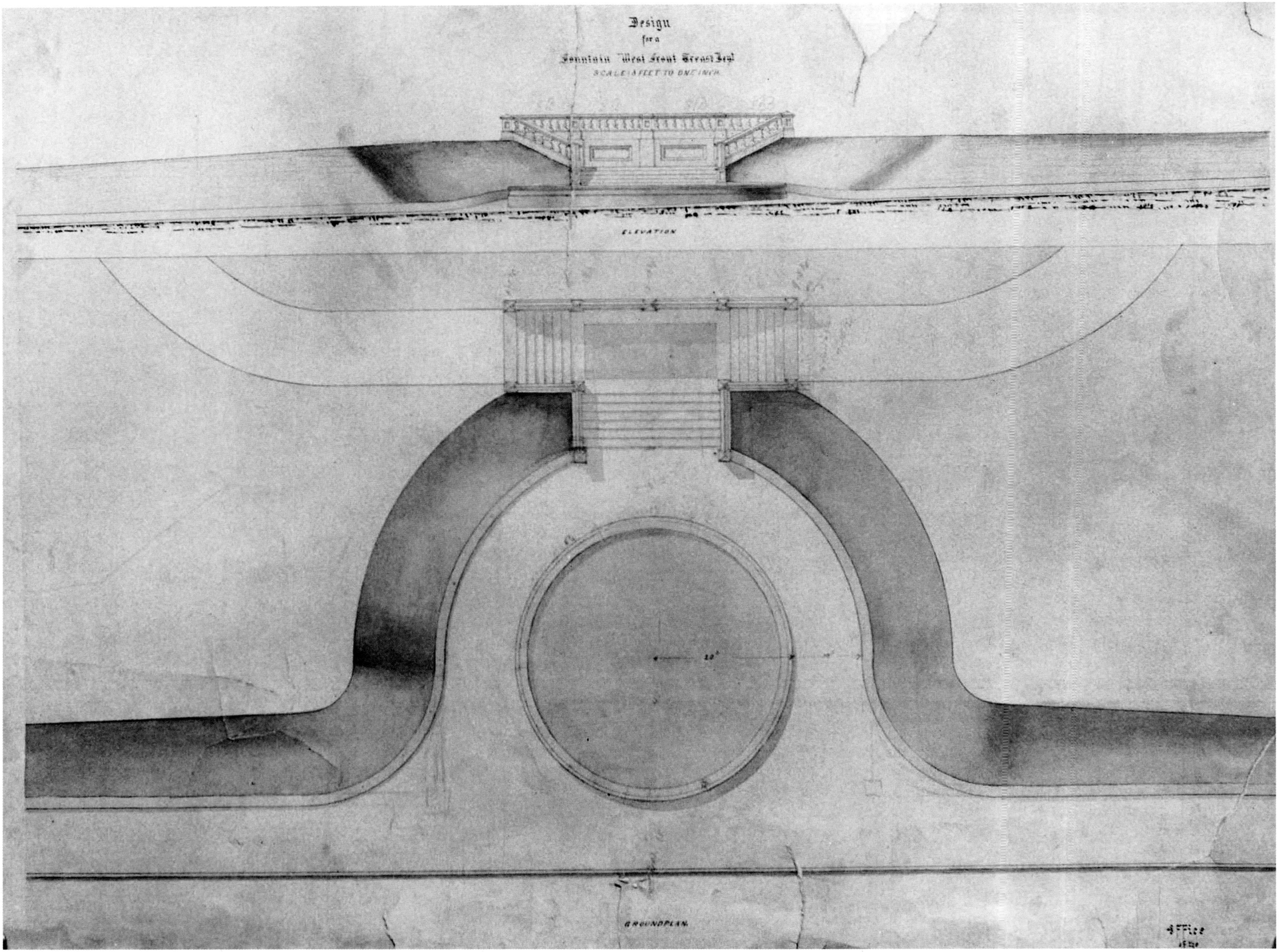
Design
for a
Fountain West Front Treas'y Dep't
ELEVATION
20'
GROUNDPLAN

Architect Alfred Mullett's original design (opposite) called for granite balustrades to flank the stairs to the East Entrance, but a low iron fence, matching that installed around the perimeter of the South Grounds, was substituted, most likely due to cost concerns. Apart from this minor alteration, however, the entire ensemble was built according to his plan. Mullett's design did not include a planting plan, but a formal parterre (pictured left in 1898) and other plantings were eventually added. The East Entrance fountain Mullett designed is seen in this photograph from 1870 (right), shooting a powerful jet of water skyward in a single stream.

the surviving West Terrace. The obvious solution would have been to rebuild the East Terrace, thereby filling in this gap and restoring the overall symmetry of the White House. However, Alfred Mullett, the architect in charge of the Treasury Building, advocated for the creation of a new East Entrance that could hold its own beside the massive structure next door. President Johnson evidently agreed and authorized him to begin work on the structure as part of the East Executive Avenue project.

Mullett created a formal East Entrance by appending a single-story colonnade onto the exposed East Facade of the mansion. It was finished in the fall of 1866. The new east entrance became the formal public entrance onto the White House landscape.

As originally developed by Mullett, visitors were first greeted by an immense 40-foot diameter granite fountain basin installed adjacent to East Executive Avenue. This fountain featured a powerful central jet that shot water skyward in a single stream. Once past the fountain, one would ascend two flights of stairs, rising to the level of the White House, and proceed along a curving walkway to the colonnaded East Entrance.

Mullett's design did not include a planting plan, and historic photographs show that the landscape featured only a grassy lawn for nearly twenty years. In 1885, however, gardeners spaded up the ground and planted a large bed of rhododendrons and azaleas where the Jefferson statue had once briefly stood. Five years later, a magazine article reported that "magnificent roses—like grown trees" also bloomed in great beds along the approach to the East Entrance. By March 1894, masses of yellow crocuses further adorned the area, and two years later fine lilacs that produced rare double-clustered blossoms were present as well. The former reflected Frances Folsom Cleveland's taste in flowers and the latter that of her successor, Ida Saxton McKinley, wife of President William McKinley.

As part of his 1902 remodeling of the White House, the architect Charles Follen McKim rebuilt the historic East Terrace according to Jefferson's original design. This reconstruction balanced the surviving West Terrace and provided a space to accommodate the increasingly large number of guests

In 1902, architect Charles Follen McKim added a new social entrance to the far end of the East Terrace. It was built around a large waiting room and included a porte cochere to shield guests from bad weather. A new horseshoe-shaped entrance drive, centered on Mullett's fountain and somewhat resembling the North Entrance Drive, provided access to this overhang.

arriving for presidential receptions and State Dinners, but it eliminated the single-story colonnade added by Mullett in 1866 and left the mansion with no formal East Entrance. McKim solved this problem by building a new social entrance onto the far end of the East Terrace. The rectangular structure recalled Mullett's earlier design through the use of paired columns arranged into colonnades.

Until the modern East Wing of the White House was built in 1942, McKim's visitor's entrance changed very little. With the onset of World War II, President Franklin Roosevelt needed additional military strategists, planners, and similar personnel stationed at the White House but lacked the requisite office space. His initial plan to build temporary housing on the South Grounds to accommodate these war workers was quickly abandoned in favor of building a permanent East Wing onto the East Terrace. Because of wartime urgency, the project proceeded at a blistering pace and was completed by year's end. Designed by Lorenzo Winslow, FDR's resident architect at the White House, the modern East Wing is a two-story structure that incorporates an expanded porte cochere leading inside the White House and out onto the Jacqueline Kennedy Garden. Despite the speed with which it was built, the East Wing neither obscured nor significantly weakened the historic character of the grounds. A few trees had to be cut down and a slight reconfiguration of the driveway was necessary, but for the most part the new building settled in comfortably and unobtrusively between the East Terrace and Mullett's East Executive Avenue fountain.

More significant changes unfolded in 1947, when the fountain and old steps were removed to enlarge the East Entrance. As a nod to history, the outline of the former fountain remains in the form of a circle inset into the sidewalk, and a surviving portion of Mullett's original iron fence now

sweeps around the open space. These changes seem to have been spurred by the need to ease traffic congestion on the White House Grounds and facilitate as many automobiles as possible around the East Entrance. Traffic jams had been a nuisance since at least the late 1920s, particularly during major events when there was not enough room to load and unload all of the guests. Although the need to preserve the surviving elements of the White House's historic landscape had been recognized by this time, some degree of change was, as always, inevitable. In the case of the East Entrance, the loss of the fountain was necessary to increase traffic flow and continue to meet the needs of the modern presidency.

During the latter part of the twentieth century, growing security concerns spurred further changes to the East Entrance. In the 1980s, a separate building was constructed south of the East Wing as a checkpoint for screening the public before a tour of the White House or its grounds. In 1983 East Executive Avenue was permanently closed to vehicular traffic and became a pedestrian-only route. Redesigned as a small, publicly accessible park, it provided some of the best glimpses onto the President's Grounds from outside the perimeter. A grass panel and new willow oak trees replaced much of the former roadbed. New round planters containing seasonal flowers and benches were also added, and pedestrians made regular use of this space. In 1986, during the presidency of Ronald Reagan, a fountain returned to the East Entrance. East Executive park, as this space came to be known, remained freely accessible until the terrorist attacks of September 11, 2001, led to its indefinite closure. At present, East Executive park remains closed to the general public. The East Entrance is still the primary visitor's entrance at the White House, but access is limited to invited guests and prescreened tour groups.

Smaller than the original, the fountain added in 1986 (above left) was removed in 2002. Mature boxwood shrubs line the steps that lead to the entrance, just beyond the fountain.

PART II

THE USES OF THE PRESIDENT'S GROUNDS

PUBLIC ACCESS *118* / PRIVATE USES BY FIRST FAMILIES *144* /
THE WHITE HOUSE GROUNDS DURING WARTIME *172*

PUBLIC ACCESS

TODAY, WASHINGTON, D.C., boasts one of the most extensive park systems in the United States. Comprising many thousands of acres managed by either the federal or the city government, the parks of the District of Columbia include not only public parks but national monuments and memorials, parkways, playgrounds, recreation centers, sports fields, amphitheaters, hiking and biking trails, urban forests, green corridors, and other open spaces. Pierre Charles L'Enfant's plan for the city also created many small circles and squares reserved for public use. These "parklets" are now ornamented with statuary, festooned with flowers, and planted with trees, shrubs, and grass. The park system bountifully serves the needs of residents and attracts visitors to the nation's capital. But this was not always the case.

Entered according to act of Congress in the Year 1850, by Robert P. Smith, in the Clerks office of the District court of the Eastern District of Pennsyl.a
VIEW OF WASHINGTON.

A bird's-eye view made c. 1850 of the U.S. Capitol building, with the Smithsonian Castle and the unfinished Washington Monument in the distance, includes the open canal that is now Constitution Avenue.

A wood engraving published in Frank Leslie's Illustrated Newspaper *c. 1874 shows the unfinished Washington Monument, with sailing boats on Tiber Creek in the foreground.*

WASHINGTON, D. C.—PRESENT CONDITION OF THE WASHINGTON MONUMENT, ON THE BASEBALL GROUND, NEAR THE WHITE HOUSE.—SKETCHED BY FRANK SCHELL.

For much of the nineteenth century, Washington's parks existed primarily on paper. As late as the 1870s, the area that is now the National Mall remained largely undeveloped, spoiled by a poorly maintained canal that was in reality a fetid, open sewer (today Constitution Avenue covers its original path). The abhorrent stench ensured that promenaders and picnickers stayed far away, and criminals and other miscreants moved in. The Washington Monument, half completed for lack of funds, stood on what was then a ragged patch of land that been the site of a Union cattle ground and slaughterhouse during the Civil War. Although there were a few green spaces like Lafayette Park, the city had no major public parks, playgrounds, or outdoor entertainment spaces. Government records state that fully two-thirds of the small circles and squares were simply "vacant and unimproved" land.

In development of parkland, Washington, D.C., lagged far behind New York, Boston, Chicago, and other major U.S. cities, a fact noted by Brevet Brigadier General Nathaniel Michler, the federal officer in charge of D.C. public lands, when he appealed to Congress in 1870: "Every important city in the country," he observed, "has acknowledged the necessity and wisdom of encouraging public places of resort for the purposes of pleasure and recreation." Congress took action, but progress was slow, and the city's modern park system did not begin to take shape until the first decades of the twentieth century. In the meantime, as people availed themselves of whatever verdant spaces they could find, the President's Grounds were a popular destination.

The President's Grounds as the Public Grounds

BY THE EVE OF THE CIVIL WAR, the presidential landscape was so fully in use by the public that it was commonly known as the "Public Grounds" and essentially functioned as de facto parkland. Public use began with Thomas Jefferson's presidency. In 1807, he exhibited two live grizzly bear cubs on the North Lawn. A gift from explorer Lieutenant Zebulon Pike Jr., who captured them on a western expedition, the cubs were carried at first by Pike's men on their saddles. When they grew larger he had a cage made for them and strapped it to a pack mule. Throughout the ordeal the cubs somehow remained meek, almost tame, a characteristic that no doubt contributed to their immense popularity with the crowds of people who came to see them outside the White House.

The December 18, 1858, edition of Harper's Weekly *newspaper published an engraving of the mansion and adjoining South Lawn and gardens with the caption "The White House and Public Grounds." The idealized scene shows only a few of the best-dressed guests strolling within a more lush setting than seen in period photographs. The message, however, was unmistakable: the White House Grounds were now first and foremost a public space.*

Men lounging around the Jefferson statue on the North Lawn, c. 1860s. With few public parks in the city at the time, the White House Grounds were inviting to the public.

The North Lawn hosted other visitors from the West as well. President Jefferson instructed Meriwether Lewis, of the famous Lewis and Clark Expedition, to encourage Native American chiefs and tribal elders to travel to Washington to meet with him, and a steady stream of delegations began arriving at the White House. While formal ceremonies took place indoors, many less formal receptions were held outside in public view. The mere presence of the Native Americans enthralled local citizens, government officials, and foreign diplomats, but the biggest thrills came when warriors performed traditional dances outside the front door.

President Jefferson also began the tradition of hosting annual Fourth of July parties at the White House. Early in the day he reviewed uniformed militia companies from the front steps of the White House, watching as they paraded, assumed various formations, and fired salutes in his honor. To the north, at the site of today's Lafayette Park, the public descended on the grassy expanse known at the time simply as "the common" to enjoy a typical country fair. Here, amid temporary tents and stalls, men played games

Robert Cruikshank fec.

President's Levee, or all Creation going to the White house, Washington.

THE GREAT CHEESE LEVEE.

The public was invited to the White House following the inauguration of President Andrew Jackson in 1829. Jackson's departure from the party made little difference. The restless crowd refused to leave until they were eventually lured outside by tubs of ice cream and whiskey punch, which served to fuel the continuing melee that ran roughshod over the White House Grounds. This engraving from the 1840s, entitled President's Levee, or All Creation Going to the White House, *depicts a Jackson-era levee.*

designed to test their skill and strength while others watched horse and dog races or bet on cockfights. Picnickers bought rugs to sit on and baskets of food and drink for toasting American independence. The merrymaking inevitably extended to the North Grounds of the White House, as Jefferson personally invited the crowd inside the mansion itself.

Although Jefferson restricted public festivities to the North Grounds, in subsequent decades access was gradually extended to the previously private South Grounds. This transition began dramatically with President Andrew Jackson's 1829 inauguration reception. More than 20,000 boisterous supporters unceremoniously pushed their way through the White House in hopes of shaking hands with the new president. Plate and china crashed to the floor, and rowdy men with muddy boots thought nothing of standing on the elegant upholstered furniture in order to get a better view of the fracas. Jackson, unable to breathe, beat a hasty retreat back to the hotel where he had been staying.

Following two terms in office, Jackson's presidency ended in much the same way as it began—with a party. On February 22, 1837, the 105th anniversary of George Washington's birthday, the public was again invited to the White House, this time to share a mammoth cheddar cheese, a gift from Jackson's supporters in Oswego County, New York. One of the leading authors of the time, Nathaniel Parker Willis, attended the event and observed that the North Lawn was "thronged with citizens of all classes, those coming away having each a small brown paper parcel and a very strong smell of cheese." Another spectator was less kind in describing the scene: "The company reminded one of Noah's ark—all sorts of animals, clean and unclean. There was quite a superabundance of the latter for the rag-a-muffins of the city had got into the gardens—thence clomb [climbed] to the terrace—and thence entered by the windows into the East Room." As they cleaned up, the White House gardeners and domestic staff might have been the only people in America happy to see the popular Jackson retiring.

Four years later, on the Fourth of July 1841, President John Tyler watched from the South Portico as a military parade stamped about on the unim-

At the end of Jackson's presidency, the public showed up to see the old general off in numbers nearly equal to those who welcomed him to Washington eight years before. On the eve of leaving office in 1837, Jackson presented the crowd with a 1,400-pound wheel of cheese that he received two years prior as a gift. The public consumed the entire cheese in only two hours. Although the crowd was noticeably better behaved than the mob that arrived to celebrate Jackson's first inauguration, both the White House and its grounds were again overrun.

Enjoying public access, visitors are seen strolling on the South Grounds in this hand-colored illustration published in the 1850s.

proved ground south of Jefferson's ha-ha wall, the future site of the Ellipse. As night dawned, a fireworks display replaced the precision marching and celebratory firing of cannon. Tyler had opened the South Grounds of the White House for the daylong festivities, and ladies and gentlemen politely strolled about, sat on the grass, and dangled their legs over the edge of the old boundary wall. According to poet and author Elizabeth Ellet, the space was virtually filled with "well dressed multitudes of either sex."

President Tyler enjoyed entertaining and evidently decided to keep the South Grounds open on a permanent basis. The space soon became a gathering place for Washington's political and social elite. As part of its new role, Tyler had the U.S. Marine Band play a free public concert on the South Grounds every week during the warm weather months, usually from May or June to late September or early October. These popular events continued, with some interruptions for war or illness and death in the presidential household, for nearly a century. Period accounts describe senators, congressmen, cabinet members, foreign ambassadors, business leaders, military officers, and other well-respected men rubbing shoulders beneath the eaves of the White House. Women in attendance dressed fashionably and kept the mood light as guests promenaded around the lawn, the very picture of genteel society. Charles Dickens, on a tour of America in 1842, attended one of the Marine Band concerts. The grounds of the White House, he said, were "pretty, and agreeable to the eye" but had the "uncomfortable air of having been made yesterday."

The public's presence on the White House Grounds became more egalitarian under President Zachary Taylor, who took office in 1849. The VIPs remained, but they had to make room for the burgeoning middle class: innkeepers, merchants, clerks, and nurses with children in tow. When D. W. Mitchell, a visiting English gentleman who spent ten years touring the United States, noticed groups of "roughly dressed" laborers lounging on the grass, he did not frown on this parity but considered the scene to be "clear unmistakable proof of the progress of people in self-government; a proof that all classes can mix freely together, each amply protected by a general regard for order and propriety and by mutual respect."

By the time of James Buchanan's presidency in the late 1850s, this democratization of the landscape had progressed to such an extent that the President's Grounds functioned as a true public park. A high point was the visit of 18-year-old Prince Albert Edward, the future King Edward VII of Great Britain. He arrived in Washington on October 3, 1860. At the White

PRESIDENT'S HOUSE,

This photograph taken from the South Lawn in 1884 captures President Chester A. Arthur on the South Portico greeting visitors. The grounds would continue to be open to the public until 1893.

House he was introduced to President Buchanan and served as the guest of honor at a fireworks display held on the South Grounds. Despite a driving rain that ruined three of the event's seven set pieces, thousands of spectators attended the show, and newspapers lauded it as a rousing success. The spectacular grand finale featured combined American and British symbols to denote the friendship between the two nations. It was reportedly the largest single fireworks piece manufactured to that time, 160 feet long and 50 feet tall.

Amazingly, the public was not shut out of the President's Grounds at the start of the Civil War. Records are spotty, but anecdotal evidence indicates that it was not until the spring of 1863 that Abraham Lincoln finally acquiesced to security concerns and closed the gates. Before this time, sightseers roamed the State Floor and the grounds, and occasionally happened to meet the president himself. From today's perspective, such an arrangement seems foolish—even dangerous—but Lincoln's objective was to foster goodwill and maintain public support for the Union war effort. It also reflects how deeply ingrained the notion of the Public Grounds at the White House had

Picnics, such as the one captured in this 1873 photograph, were one of many ways the public was able to enjoy access to the South Lawn in the late nineteenth century.

become by this time.

With the close of the Civil War, the presidential landscape reopened to the public. African Americans, many of them former slaves, also began to make regular use of the space. The President's Grounds, now Public Grounds in the truest sense, were open to all classes of people and all races as well. This egalitarianism was perhaps most evident during the annual White House Easter Egg Roll. Begun in 1878 under President Rutherford Hayes, the event grew from hundreds to thousands of participants, including "all colors and all ages, from the toddler of three to the school lass of thirteen, and apparently all conditions of life."

By the 1880s and 1890s the daily crowds on the President's Grounds began to seriously impinge upon the needs of the first family for private outdoor space. Foreign visitors to Washington were often amazed that such a duality existed at all, as demonstrated by an amusing story from the time of President Benjamin Harrison. A visiting Englishman was astonished that ordinary people could wander the White House Grounds. He told his American friend, "It's wonderful, and make no mistake. Why, they let you do just as you please. Do you know, I think that if a fellow wanted to he could go in there and roll over on the lawn and there wouldn't be a person who would think of stopping him?" Then, after a minute's deliberation, he did just that. Laying down flat on his back in the middle of the crowd, the man "rolled over three times, slowly and deliberately. Then he got up and walked out of the grounds," as though nothing unusual had happened. If he had acted that way in any of the European capitals, said the Englishman, "he would have been locked up as a dangerous character."

The era of President's Grounds as Public Grounds effectively ended in 1893, when President and Mrs. Grover Cleveland locked the gates leading to the South Grounds to protect their young children. The North Grounds remained open for a while, and the South Grounds allowed public visits on a few brief occasions until the entire White House landscape was permanently closed during the Coolidge administration. The weekly Marine Band concerts outlived the general closure of the grounds by a few years, and continued until 1929. The Easter Egg Roll still takes place every year, but is now a ticketed event.

The Weekly U.S. Marine Band Concerts

THE U.S. MARINE BAND, the oldest continuously operating musical organization in America, first performed at the White House for President John Adams's New Year's Reception on January 1, 1801, making its association with the mansion nearly as old as the building itself. The band has long been known as "The President's Own," and it performs at receptions, State Dinners, formal ceremonies, and diplomatic events. Most Americans are accustomed to hearing "Hail to the Chief" whenever the president makes an official appearance, but a lesser-known and largely forgotten tradition was the weekly public concerts on the South Grounds.

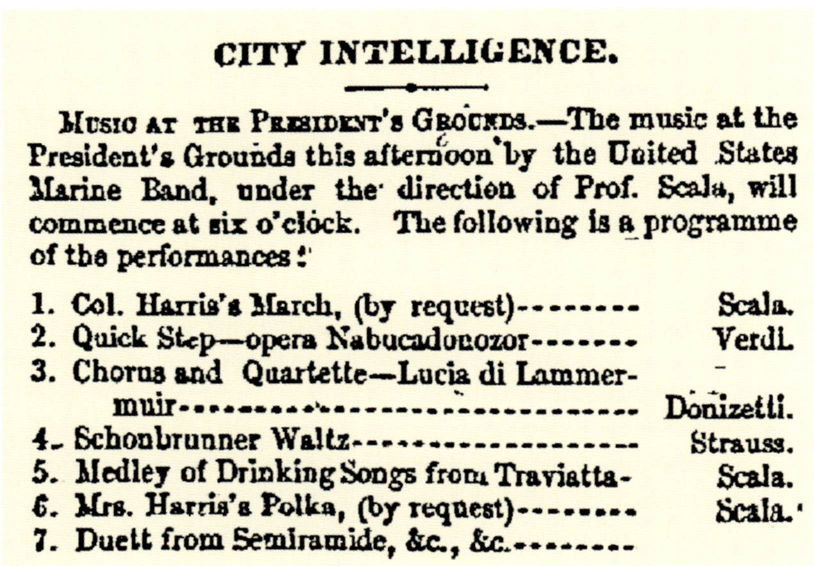

CITY INTELLIGENCE.

MUSIC AT THE PRESIDENT'S GROUNDS.—The music at the President's Grounds this afternoon by the United States Marine Band, under the direction of Prof. Scala, will commence at six o'clock. The following is a programme of the performances:

1. Col. Harris's March, (by request)--------	Scala.
2. Quick Step—opera Nabucadonozor-------	Verdi.
3. Chorus and Quartette—Lucia di Lammermuir------------------------------	Donizetti.
4. Schonbrunner Waltz------------------	Strauss.
5. Medley of Drinking Songs from Traviatta-	Scala.
6. Mrs. Harris's Polka, (by request)--------	Scala.
7. Duett from Semiramide, &c., &c.--------	

The September 7, 1859, Evening Star *carried an advertisement for the program for that evening's Marine Band performance on the President's Grounds.*

On July 21, 1861, the Marine Band performed under a flag-crowned pavillion built on the South Grounds. President Lincoln and his guest Prince Napoleon are seen standing on the South Portico.

Held spring to fall, these concerts began in the early 1840s under President John Tyler and continued each year until President Herbert Hoover discontinued them nearly a century later. Things got off to a rocky start as a sudden storm ended the first outdoor concert shortly after it began and sent Tyler and his guests scrambling into the White House to keep from getting drenched. Thankfully, the weather cooperated the following week, and people turned out in great numbers. The concerts quickly became so popular that they were moved to Saturday afternoons from Wednesdays, trading days with a similar weekly performance by the Marine Band outside the U.S. Capitol Building.

While Tyler and his successor, President James K. Polk, usually remained on the South Portico during the performances, the next president, Zachary Taylor, elected as a common-man candidate, walked about the South Grounds, shaking hands and patting backs as he went. One period account noted how he mingled "out among the crowd" of three to four hundred people and seemed "kind and agreeable, both in appearance and manner."

PRINCE NAPOLEON VISITING PRESIDENT LINCOLN AT THE WHITE HOUSE, WASHINGTON, D. C.—SCENE ON THE PALCONY DURING THE PERFORMANCE BY THE BAND.—From a [illegible]tch by our Special Artist accompanying Gen. McClellan's Command.

An illustration made for Harper's Weekly *depicts the 1868 Fourth of July celebration on the White House Grounds. The platform built during the Lincoln administration is still in use by the Marine Band. The large, well-dressed crowd includes a number of men in military uniform. After the Civil War, Francis Scala, leader of the U.S. Marine Band from 1855 to 1871, noted in his memoirs that both President Andrew Johnson and his wife, First Lady Eliza Johnson, admired and supported the band and its musicians. Although illness kept Mrs. Johnson confined to her rooms on the southwest corner of the Second Floor, she could see and hear the Marine Band through her window.*

He particularly "delighted with the children who leapt about so joyously and so free from care, or seated themselves on the green sward." Newspaper announcements of programs reveal that the Marine Band played both classical and popular musical selections. In August 1859 the program included waltzes, quicksteps, polkas, and operatic adaptations.

Even as the Union was dissolving, President James Buchanan continued the Saturday concerts as a means of drawing people together and maintaining a sense of national unity. Lincoln did so for much the same purpose, and after the fighting began he added patriotic songs to the band's repertoire. During the Civil War the South Grounds were routinely "thronged with visitors . . . enjoying both spectacle and music." Lincoln had a wooden platform constructed for the Marine Band, complete with a blue-and-white striped canvas that shielded the musicians from the sun. In addition to stirring up pro-Union sentiment, the concerts were important to Lincoln because they offered brief respites from his heavy duties. Yet whenever he appeared the music soon gave way to applause and calls for a speech, so after waving to the crowd from the South Portico he invariably stepped back indoors. He was once heard to remark that he wished his supporters would simply "let me sit out there quietly, and enjoy the music."

The platform built by Lincoln remained in place after Andrew Johnson became president, and he and his wife Eliza supported the band and its musicians. Too frail to be in public, she enjoyed the concerts from her bedroom window. The next president was a former general, and Ulysses S. Grant had a preference for military music and musicians. As the weekly Marine Band concerts grew in popularity, Grant's expansion of the South Grounds accommodated larger and larger crowds. His successor, President Rutherford Hayes, was more traditional in his tastes, and he and his wife Lucy Webb Hayes preferred religious hymns, musicales, and choirs over the popular music of the day. It seems that the pure entertainment value of these weekly concerts may have ebbed during their time at the White House. Further, the remainder of the 1881 season was canceled after an assassin shot President James A. Garfield on July 2. The concerts regained their popularity, however,

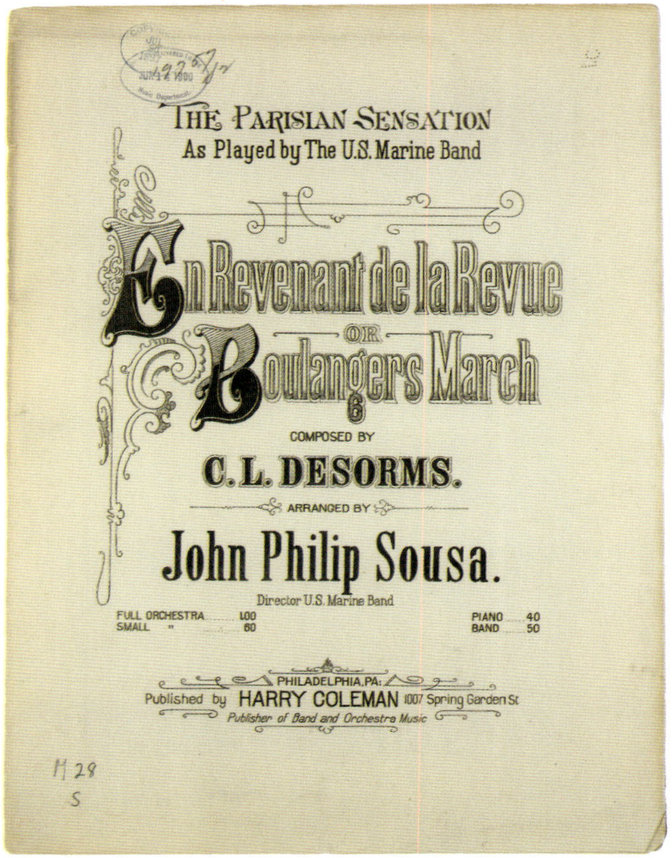

The cover to the sheet music for Boulangers March, *arranged by John Philip Sousa, names the Marine Band. The catchy tune was first played by the Marine Band during the Cleveland administration and became a regular part of the band's repertoire.*

By the 1920s, the Marine Band regularly performed its weekly concerts from an elevated platform on the South Lawn while the audience sat on the grass in an orderly fashion, as depicted in this photograph taken on July 16, 1921, during the administration of Warren G. Harding.

under President Chester A. Arthur, who revived the long-dormant custom of mingling with the people out on the lawn.

The Marine Band played a memorable concert on June 20, 1885, during President Grover Cleveland's first term, when a large crowd gathered to hear the musical score from Gilbert and Sullivan's comic opera *The Mikado*. Held a mere three months after its world premiere on the London stage, and before any official productions opened in the United States, this concert on the South Grounds previewed a work that would soon take the country by storm. Two years later, the Marine Band gave one of the first U.S. performances of the French piece "En Revenant de la Revue," known in America as "Boulangers March." The audience listened to the catchy tune with "great pleasure," and the march became a regular part of the band's repertoire.

The conductor of the Marine Band during these years was the famed composer John Philip Sousa, who had served a seven-year musical apprenticeship with the band and then returned to lead it from 1880 to 1892. He often composed his own arrangements and pleased audiences with impromptu selections. For example, once he closed a concert with "Buffalo Gals Are You Coming Out To-night?," a reference to First Lady Frances Cleveland's hometown.

Mrs. Cleveland especially enjoyed the work of German composer Richard Wagner, and thousands of Americans probably heard his stirring music for the first time when the Marine Band adapted his orchestral scores for its Saturday concerts. President Cleveland, however, preferred light music over Wagner's loud, deep arrangements. During one such performance on the South Grounds, Sousa recalled seeing the president stand up and hold onto a nearby window "as though he were afraid it would get away from him" while "his pretty young wife sat down facing him, enthralled" by the soaring tune.

During the last decades of the nineteenth century all America was caught up in a "band craze." According to the White House music historian Elise Kirk, "Not only were military bands all the rage, but so were industrial bands, Salvation Army bands, circus bands, and school bands. . . . Hardly an event could be properly observed without a band, and every village or

The Marine Band continues to perform on the grounds today along with other bands, such as the United States Army Old Guard Fife and Drum Corps, seen here during the State Arrival Ceremony for the prime minister of Japan, Shinzō Abe, and Mrs. Akie Abe on April 28, 2015.

The Marine Band played patriotic and popular tunes for the visiting public during the 2016 Spring Garden Tour.

town seemed to have one." Not surprisingly, the popularity of the weekly Marine Band concerts peaked during this time. Moreover, the band itself "linked the presidential family with the nation like no other social or cultural phenomenon of the period."

This effect was most pronounced during the tenure of President William McKinley, when up to 5,000 people attended the Saturday performances. Such large crowds led one reviewer from the *Washington Evening Star* to quip that most people simply attended automatically, "having been inoculated with it in infancy." Their elegance was notable. A newspaper account described "summer girls" in bewitching gowns, all "visions of frills, laces and ribbons," promenading beneath the trees, each with a "summer boy" dressed impeccably in white, from his hat to his canvas shoes. Performances ended with "Hail, Columbia," then the unofficial national anthem. Toward the end of the 1902 season one concert played "Nearer My God to Thee," McKinley's favorite hymn. It was the one-year anniversary of his assassination, and heads were bowed in grief.

The Marine Band concerts continued throughout the administrations of Theodore Roosevelt and William Howard Taft. During the Wilson administration, after the United States entered World War I, the concerts were relocated to the Ellipse because of security concerns. They returned to the South Grounds for the 1919 season, but in 1922 some performances were canceled owing to First Lady Florence Harding's illness. They were suspended the next year after President Warren G. Harding's death. President Calvin Coolidge temporarily revived the tradition after ascending to office, but by then the concerts' popularity had evidently begun to wane. President Herbert Hoover finally discontinued the weekly Marine Band concerts in 1929.

Although the weekly public concerts have not been held for nearly a century, the Marine Band continues to perform on the White House Grounds to this day. It often provides the musical accompaniment for the official ceremonies as well as private gatherings. At times additional U.S. military bands also participate in these events.

The Annual White House Easter Egg Roll

EXACTLY WHEN THE ANNUAL CUSTOM of Easter egg rolling began in Washington is uncertain, but it is likely that the events were first hosted on the U.S. Capitol Grounds in the early nineteenth century. Every year on the Monday after Easter, children gathered to roll hard-boiled, brightly dyed eggs down the slope of Capitol Hill, laughing as they tumbled down after them. By the 1870s, the event was a bona fide institution, with several hundred children participating. But each year the space was a little worse for the wear, and Congress had to pay for repairs to the landscaping. The 1876 event proved especially damaging, and shortly thereafter Congress passed a law forbidding the Capitol Grounds from being used as a children's playground, thereby effectively outlawing the Easter egg rolling event. This legislation was to be enforced in 1877, but a severe rain kept the children indoors on Easter Monday and staved off a confrontation until the next year.

NO EGG-ROLLING ON THE CAPITOL GROUNDS. By an act of Congress approved April 29th, 1876, it is made the "duty of the Capitol police hereafter to prevent any portion of the Capitol grounds and terraces from being used as playgrounds or otherwise, so far as may be necessary to protect the public property, turf and grass from destruction or injury." This law was passed to prevent the practice which had grown up in previous years of children rolling Easter eggs on the Capitol grounds and terraces.

This small notice published in the Evening Star *in 1878 alerted the public that egg rolling would no longer be allowed on the Capitol Grounds.*

Two children take a break from the egg rolling on the South Lawn to pose for this photograph by Frances Benjamin Johnston in 1889. During these events attention was not paid to race or class distinctions. As with any other day, all were welcome. Thus the rolling of Easter eggs at the White House has always been a thoroughly democratic event, from Hayes's time on up to the present day.

An announcement appeared in the morning papers on Easter Sunday 1878 informing prospective egg rollers to steer clear of the Capitol. Later that day a small group of children happened upon President Rutherford B. Hayes while he was out on his daily walk and asked him if they might be allowed to use the South Lawn of the White House instead. Hayes was perplexed by the polite request. Since his first Easter as president had been washed out by the previous year's rainstorm, he knew nothing of the tradition. He did, however, bring the matter up with his aides when he returned to the mansion. They told him not only of the event's history but of similar events outside the White House under Presidents Andrew Johnson and Ulysses S. Grant. These had all been small family affairs, but now Hayes was only too happy to make egg rolling on Easter Monday a public event. He issued an official order stating that any and all children would be welcome to come and roll their eggs on the President's Grounds. Word spread quickly, and some two hundred children showed up on April 22, 1878, the first of the annual Easter Egg Rolls at the White House. The "Public Grounds" now had a yearly holiday all its own.

Children gather eggs on the South Lawn in 1889, the first year the U.S. Marine Band played at the event.

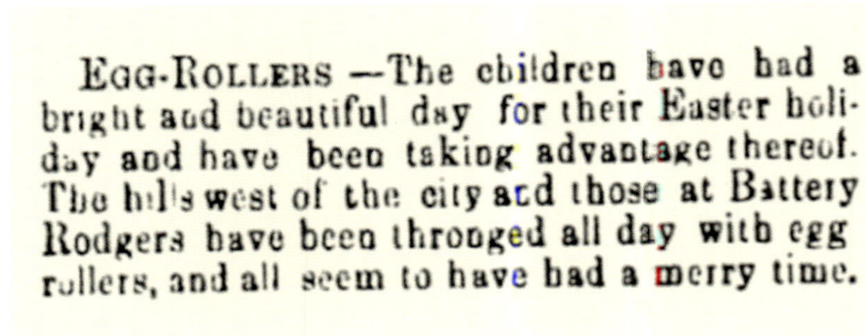

EGG-ROLLERS.—The children have had a bright and beautiful day for their Easter holiday and have been taking advantage thereof. The hills west of the city and those at Battery Rodgers have been thronged all day with egg rollers, and all seem to have had a merry time.

An excerpt from the Alexandria Gazette *described egg rolling on April 22, 1878, the day of the first annual Easter Egg Rolls at the White House.*

A crowd gathers to watch as a group of boys run down the sloping South Lawn in pursuit of their eggs during the 1926 White House Easter Egg Roll.

The earthen mounds on the South Grounds made for good rolling, both of eggs and of children, and the celebration was a resounding success. The *Washington Evening Star* reported that the whole lot "laughed, yelled, and played," including several girls who, upon following after their eggs, "were totally regardless of the extent of striped stockings displayed." Reporting on the 1890 White House Easter Egg Roll, the *Baltimore Sun* showed just how popular the gathering had become:

> There were all sorts and conditions of children—big children and little children, thin children and fat children, blondes and brunettes, white children and colored children, children in caps and pinafores, children in reefer jackets and Tam O'Shanter caps, in Gretchen gowns and kid shoes, in calico gowns and the barest excuses for shoes, rich and poor, under the care of nurses or roaming about at their own sweet will. But in one respect they all agreed—that all had a merry time together, without regard to dissimilarity of age or condition. . . . With their parents and nurses they completely filled the grounds, and it as estimated that at times during the day there were as many as 50,000 people collected there.

In an era when racial segregation was the practice and even the law in much of the country, and when the rise of the industrialist "robber barons" had widened the gulf between the rich and the poor in American society, the White House welcome given all children and their parents is notable.

Children arriving for the Easter Egg Roll in 1889 found that a wooden bandstand had been erected on the South Lawn, the first time since Lincoln's old platform was removed in 1878. Soon the Marine Band arrived and played "Hail to the Chief" as President Benjamin Harrison stepped out onto the South Portico, holding his young grandson by the hand. Thunderous applause erupted, and then the band performed a concert for the children's

The U.S. Air Force Band performs during the Easter Egg Roll in 1966 during the Lyndon B. Johnson administration (above left), and President and Mrs. Ronald Reagan join children at the starting line of the egg rolling competition, 1982 (above right).

enjoyment. The selections from that first program were not recorded, but the following year the *Philadelphia Enquirer* listed a few of the "familiar airs" that added to the day's enjoyment: "Suwanee River," "Where Did You Get That Hat?," and the seemingly out of place "Listen to My Tale of Woe."

Even President and Mrs. Grover Cleveland, who effectively ended the public park era on the President's Grounds, enjoyed the Easter Egg Roll. The first lady watched the fun from atop the South Portico and even "romped and played" like a schoolgirl on the lawn below. President Cleveland tried his talents at "egg picking," a game in which players took turns knocking their eggs together. The first egg to crack lost the match and was usually eaten by the winning player. Other popular games of the era included "egg ball," "toss and catch," and "egg croquet"—which all seem self-explanatory and attest that in those days the White House Easter Egg Roll could be smelled almost as easily as it was seen or heard. When the children were thoroughly exhausted, and apparently thoroughly messy as well, the president invited the whole troop into the White House and shook their little hands as if he were greeting foreign diplomats, before retiring for the evening.

In 1905 adults were banned from the Easter Egg Roll unless accompanied by a child. With the immensely popular Theodore Roosevelt and his family in the White House, attendance at the yearly event had grown to the point that there was simply no space on the South Grounds for grown-ups without children. Those denied entry tried to "borrow" children outside the gates, while opportunistic youngsters quickly realized they could earn some pocket change from the situation. One enterprising boy made a kiddie fortune escorting adults onto the grounds. Charging a dime a head, he would vouch that the strangers in his company were in fact his parents. Then, after the adults were safely inside, he would slip back past the guard, find a new "mom and dad," and start the ruse all over again.

In 1917, the Easter Egg Roll was temporarily relocated to the Washington Monument Grounds because of increased security measures during World War I. The next year, in accordance with a push to save some 60 million eggs for the war effort, the nation observed an "eggless Easter" and the annual tradition at the White House was suspended. The children and their eggs were also not permitted onto the President's Grounds in 1919, despite the

fact that the war was over.

When the White House Easter Egg Roll finally resumed under President Warren G. Harding, it proved more popular than ever. Soaring attendance necessitated restricting the event to children 10 years of age or younger, accompanied by a parent or other adult chaperone. Even so, a capacity crowd of 48,105 was admitted in 1926. One year later, 30,023 people were in attendance on the White House Grounds, with a further 17,000 turning out at ancillary sites within Rock Creek Park and on the Washington Monument Grounds.

During World War II, the Easter Egg Roll was once again temporarily suspended or relocated, and once more during President Harry Truman's renovation of the White House in 1948–52. On April 6, 1953, during the Eisenhower administration, the event returned to the White House Grounds for the first time since 1941. Despite chilly weather and an unusually small crowd, Ike also turned out for the Easter Egg Roll, the last presidential appearance until President Gerald Ford and First Lady Betty Ford hosted the annual tradition in 1976. Five years later, in 1981, President and Mrs. Ronald Reagan introduced the first commemorative White House Easter Eggs, decorated wooden eggs that attendees could take home with them as souvenirs

In 2003, President George W. Bush closed the event to the public and designated it a special day for military families. An estimated 12,000 people attended, all active-duty and reserve service members and their families. Lynne Cheney, wife of Vice President Richard Cheney, spoke directly to the children, telling them, "We think your moms and dads are terrific," and "we are proud of all the men and women who serve our country, who keep our country safe." Interviewed servicemen and servicewomen returned the sentiment, saying that the were "honored to be there" and that the day was a "wonderful opportunity."

The following year, the White House Easter Egg Roll reopened to the general public. However, the yearly celebration operated as a ticketed event, as the number of would-be guests was far too many to be safely accommodated on the South Lawn. Since 2009, tickets have been made available through an online lottery system. More than 200,000 tickets are requested each year, but attendance is limited to about 35,000 people.

President and Mrs. George W. Bush watch as the races begin at the 2006 Easter Egg Roll (above left), and children rush through the open gate in search of eggs in 2015 during the Barack Obama administration (above right).

PRIVATE USES BY FIRST FAMILIES

IN ADDITION to events open to the public such as Marine Band concerts and annual Easter Egg Rolls, first families have long enjoyed the South Grounds for their private use. This is the "pleasure ground" that Benjamin Henry Latrobe had envisioned for the president, and although the public has often encroached on the first family's need for privacy, the beautiful greensward has also long served as a place for presidential recreation and refreshment. The South Grounds are a lovely setting for garden parties, entertainments, and private outdoor concerts. Many presidents have enhanced the grounds with the pools and courts and courses that give them, their families, and invited guests a chance to play sports, relax, and unwind. And children's play and games—the running and the laughter and the fun—have long made the South Grounds much like anyone's backyard.

Parties on the South Grounds: "Almost into Fairyland"

ONE OF THE EARLIEST and most unusual of the private events held on the South Grounds took place on the South Lawn on July 28, 1866, during the Andrew Johnson administration. Advertised as a "Grand Gymnastic Exhibition and Instrumental Concert," the Marine Band provided the musical accompaniment for a troupe of traveling athletes and acrobats who no doubt put on an amazing show. Unfortunately, the details of their performance are lost to history.

First Lady Edith Roosevelt introduced the garden party concept to the White House in 1902. This stereograph captures a party held in 1905. Washington society soon followed suit.

Private outdoor entertaining was not commonplace, however, until the early twentieth century when First Lady Edith Roosevelt began the custom of hosting such events. She had held garden parties for years at the Roosevelts' home, Sagamore Hill, in Oyster Bay, New York, so it was only natural that she should bring the practice to Washington. In anticipation, newspapers announced that her inaugural garden party of May 8, 1902, would be "the first time in the history of the White House that the mistress of the mansion has given this kind of function" and speculated that before long all of Washington high society would soon follow suit.

The papers later described the gathering as a "beautiful fete," with several hundred invited guests in attendance, all members of the "smart set." The South Lawn served as a grand court, on either side of which two flag-draped army tents were set up for serving punches and ices. The first lady received her guests at the foot of the South Portico, standing on a Persian carpet spread out on the grass beneath the boughs of an immense elm. She wore a "charmingly simple" white organdy gown, made over white taffeta and trimmed with ruffles and lace, and a lace hat adorned with pink roses. Alice Roosevelt, President Theodore Roosevelt's eldest child and the only one with his first wife, joined her stepmother in the receiving line and shone in a pink organdy dress and white-laced hat trimmed in plumes. The president, in his riding costume, bowed in a courtly manner to the diplomats and greeted each of his friends with a well-turned phrase. The Marine Band provided the music, and when it played "Musicians in the Forest" several members obligingly relocated themselves among the trees, where their instruments echoed those of their bandmates. One attendee exclaimed that the serene scene "made a picture worthy of the brush of one of the great artists of the Watteau type." Proving that imitation is the sincerest form of flattery, the fashionable ladies of Washington soon began sending out invitations for their own garden parties.

Edith Roosevelt's garden parties were not the only innovation she brought to White House entertaining. Shortly before their time in the mansion ended, the Roosevelts arranged for performances of children's musical plays on the South Grounds. About 4,000 people attended these events, held on October 16 and 17, 1908. Adults paid an admission fee of $1 and children

25¢, and the entire $2,000 in proceeds was donated to the Washington Playgrounds Association. Under the direction of Ben Greet, a thirty-year stage veteran described as "one of the greatest living authorities on English drama," local children joined professional actors in presenting two morality tales by Nathaniel Hawthorne, *Midas and the Golden Touch* and *Pandora and the Box of Mischief*. The children and adult actors wore costumes and acted on an improvised stage. The shrubbery on the grounds served as the curtain, with the performers making their entrances and exits through dense foliage. This idyllic setting coupled with the tender age of the performers greatly impressed the audience, and President Roosevelt went "backstage" into the White House corridor that served as the dressing room to congratulate the children on their success and shake their little hands.

First Lady Helen Taft loved garden parties, and she held at least four every year that she was in the White House. The Tafts also continued to host theatrical performances. On June 16–17, 1910, the Coburn Shakespearean Players, a vaunted traveling theater troupe, performed *As You Like It* and *Twelfth Night* in a temporary amphitheater on the South Lawn. It was the first time that Shakespeare had been enacted at the White House, and the press lauded the productions as a "great step in the revival of pageantry" in America. Dozens of local schoolchildren were among the many guests, and the Washington Playgrounds Association was again the beneficiary of the ticket proceeds. The audience was so transfixed by the performance that they remained in their seats and waited out a brief summer downpour.

The Tafts' most spectacular event was their silver wedding anniversary, on June 19, 1911, a celebration that culminated with a gala garden party under the stars—except that the stars could barely be seen that night, so aglow were the South Grounds with incandescent light. Thousands of Chinese paper lanterns hung suspended from the trees, casting shapes and shadows as they swung in the breeze. Light strands in red, white, and blue were hung in swags from the tops of the largest oaks and elms, down toward the ground. Smaller lights wound around the trunks of pines and cedars.

Onlookers packed around the iron perimeter fence twenty-deep to see the show of light and hoping to get a glimpse of some of the five thousand invited guests who had to push through the crowds on East Executive Avenue to get to the entrance gates. Most had managed to make it onto the grounds by 9:00 p.m., when President and Mrs. Taft appeared, arm-in-arm, to the familiar strains of Felix Mendelssohn's "Wedding March." The presidential couple proceeded down the steps of the South Portico, over which an American flag made of red, white, and blue light bulbs shimmered and waved, and then marched to the center of the South Grounds, where they took their positions beneath the boughs of two trees joined by an electric light flashing "1886–1911." For hours they stood at the head of a long receiving line, shaking hands and kissing cheeks as a steady stream of awestruck family and friends came to congratulate them.

This party, while a dazzling spectacle, was also an informal affair. The guests wandered at will, and as one newspaperman put it, "Every corner of the mansion had its own particular light." The globes atop the east and west terraces were covered with dark red paper, and the resultant glow, caused them, in his opinion, to resemble "monster poppies." An immense spotlight, mounted atop the Treasury Building, shone upon a new American flag fluttering at the summit of the White House roof. Near this flag was the largest light of all, a navy searchlight borrowed from the dispatch ship *Dolphin*. The *Dolphin*'s crew, resplendent in their crisp, white uniforms, realigned the searchlight every so often, but most of the time they aimed the beam directly at the South Fountain. Passing through several thousand red, white, and blue lights installed on the rear of the South Portico, this beam caused the water sprays to twinkle with all the hues of the rainbow. It also brought the Washington Monument into sharp relief against the dark sky.

Light refreshments were available in tents scattered throughout the groves on the South Grounds, and the presents to the Tafts, which numbered in the hundreds and were worth in the thousands, piled up over the course of the night. Music came from the Marine Band, half concealed in the greenery, and also the U.S. Engineers Band, stationed on the South Portico.

At midnight Helen Taft retired to her bedchamber. Although exhausted, she was much too excited to sleep. Instead she turned out the lights, opened her window, and watched the party continue on the grounds below. At 1:00 a.m. the conductor of the Engineers Band asked the president if he should strike up "Home Sweet Home," a cue that the night would soon be coming to a close. Taft's answer was an emphatic "no." He was having far too much fun mixing informally with his guests, a simple pleasure that is rarely possible for the president of the United States. The music finally stopped an hour later, but even then President Taft was not quite ready to bring the enchanting evening to a close. Instead, he seated himself on the South Portico, where he remained until the last guest had gone. Alone in the early predawn hours, he decided that the following evening would be every bit as magical as the one that had just passed. The vast throng of people that had spent the better part of the night looking in from beyond the fence had not gone unnoticed.

President Taft ordered that the White House Grounds be opened the

Viewed from the Treasury Department, a spotlight shines into the large crowd outside the East Entrance of the White House during President and Mrs. Taft's silver wedding anniversary celebration on June 19, 1911. A correspondent from the Trenton Evening Times *ably summed up the scene: "With the weather absolutely perfect, the White House grounds were transformed almost into fairyland."*

During President Franklin D. Roosevelt's fourth and final inaugural ceremony in 1945 attendance inside the White House fence was limited to 5,500 people. They were separated into two sections, with perhaps a few hundred additional members of the diplomatic corps, photographers, reporters, and broadcasters also on hand.

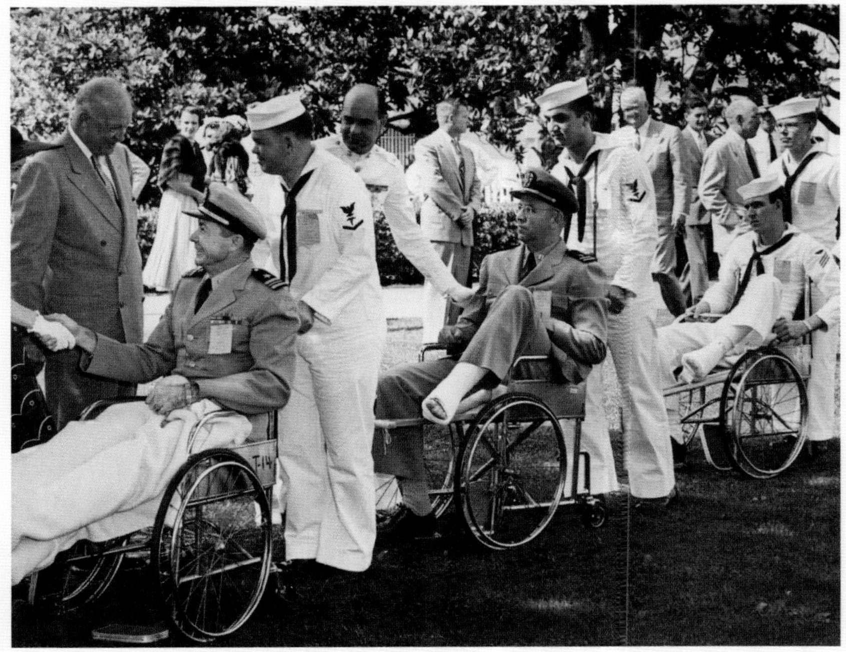

President and Mrs. Dwight D. Eisenhower greet their guests during a White House garden party held in honor of wounded U.S. soldiers on May 27, 1953. The military camaraderie was evident as Eisenhower happily chatted and mingled with the soldiers.

next night to the general public for three hours, from 8:00 to 11:00 p.m., so that everyone could enjoy the fairyland of varicolored lights emblazoning the night sky. That evening Mrs. Taft, her daughter, and her sister sat on the South Portico and watched as people strolled about with wide-eyed wonder. President Taft did not make an appearance, but a letter from the artist Francis Millet, who was among those at the public reception, ensured that he knew his kind gesture was well-received:

> A happier company of people I never saw and I am sure you would have felt as I do now, if you had heard them talk and had seen their faces. I hope some of them will write to thank you for the pleasure you gave them. If they do not it is because they are shy and not because they do not recognize the sympathetic interest, which prompted you. . . . I was so touched by the spirit of the evening . . . of what you have so gracefully done to win the hearts of those who accepted your hospitality.

First Lady Ellen Wilson is known to have hosted only one garden party at the White House before her death in August 1914. Then it was wartime, and President Woodrow Wilson's second wife, Edith Wilson, held the first-ever outdoor reception for wounded soldiers. Thirty years later, President Dwight Eisenhower and First Lady Mamie Eisenhower did the same for wounded World War II servicemen.

Franklin D. Roosevelt, president during World War II, largely stayed off the grounds, owing to his paralysis, which was carefully concealed. Garden parties were eliminated from the White House social schedule after the United States entered the war. But one event on the President's Grounds from these years is notable: the presidential inauguration of January 20, 1945, when Roosevelt was sworn in for a record fourth term as president. By this time his health was so precarious that it was deemed inadvisable for him to take the customary Oath of Office on the steps of the U.S. Capitol. Instead, an abbreviated ceremony was held on the South Lawn, with the president taking the oath on the South Portico and then delivering a short speech. For the 5,500 who were present, it was the closest that many would ever come to the man who guided them through the dark days of the Great Depression and World War II, and they saw him as the strong, resilient leader they had come to love. In truth, Roosevelt was an extremely sick man, and was dead less than three months later.

Entertainment at President John F. Kennedy's White House was legendary. Frequent after-dinner musicales prompted the creation of the Concert for Young People program in late August 1961. Vocalists and musicians came to perform for the first family and their guests, often in a temporary band shell put up on risers supplied by the Department of the Interior. In the summer of 1963, torrential rains forced a performance of Wolfgang Amadeus Mozart's *The Magic Flute*, to be moved indoors, spread out between the East Room and the Entrance Hall, with the conductor stationed in the doorway between them. Such galas had not been held outdoors at the White House since the Tafts celebrated their silver wedding anniversary fifty years earlier, and the positive public perception of these cultural events helped to build the "Camelot" mythos that remains an indelible aspect of President Kennedy's legacy.

The Concert for Young People program began in August 1961 and continued throughout the Kennedy administration. Outdoor performances were frequently staged on a temporary band shell erected by the Department of the Interior. This concert by the Greater Boston Youth Symphony Orchestra and the Breckenridge Boys Choir was held on April 16, 1962.

Under President Lyndon Johnson, the grounds hosted the White House Festival of the Arts on June 14, 1965. Designed by the president's cultural adviser, Princeton University historian Eric Goldman, the festival featured

In September 1967, the Johnson "Country Fair" brought a small amusement park to the White House Grounds, and also featured clowns, games, farm animals, and pony rides.

music, drama, poetry, painting, and sculpture and drew from every major institution in Washington. Performers included dancer Gene Kelly, orchestra leader Duke Ellington, singers Roberta Peters and Marian Anderson, writers Mark Van Doren, John Hersey, and Saul Bellow, and poet Robert Lowell. By far the largest outdoor entertainment hosted by the Johnsons was the 1967 "Country Fair," an emulation of the festivals held each summer or fall across the nation. Thirty years later, President William Clinton and First Lady Hillary Clinton reprised the experience by hosting an "old-fashioned carnival" for children.

On May 24, 1973, President Richard Nixon and First Lady Patricia Nixon held a dinner on the White House Grounds honoring 591 American soldiers who had been held as prisoners of war in Vietnam. The event also marked the Nixons' thirty-third wedding anniversary, and the gala, known simply as the "P.O.W. Dinner," was the largest formal dinner the White House had seen up till that time. An immense yellow-and-white striped tent protected the ensemble from a sudden rainstorm and allowed the evening, which alternated lighthearted music and dancing with stirringly emotional toasts and speeches, to stretch on almost until dawn.

In July 1976, President and Mrs. Gerald Ford hosted Queen Elizabeth II and Prince Philip of Great Britain at a formal State Dinner, held in the Rose Garden in a tent, to commemorate the two-hundredth anniversary of U.S. independence. Many different ideas were suggested for the after-dinner entertainment, but the proposal that most appealed to the first lady was intended to recognize the queen's well-known affection for horses: a rodeo was to be staged on the South Lawn, complete with bronco-busting, bull riding, calf roping, and steer wrestling. A selection of country music would replace the customary orchestral suites. In the end, more conventional tastes won out—just as well since it rained that evening.

But for much of the Ford presidency the country was mired in an economic downturn, and the Fords often entertained with "brown bag" lunches

In May 1973, President and Mrs. Richard Nixon hosted what was up to that point the largest formal dinner ever held on the White House Grounds. The event honored 591 soldiers who had been held as prisoners of war in Vietnam.

eaten picnic-style on the South Grounds as an example of financial frugality. President Jimmy Carter continued these thrifty ways. A lover of classical music, Carter held concerts and musicales at the White House throughout the year, moving them outdoors in the summer months, where the maximum number of guests could be accommodated for the minimum cost. To further control expenses, the menus for most of these events were informal, featuring barbeque, fried chicken, and hamburgers and iced tea and coffee as beverages. Although motivated largely out of economic necessity, the nature of these events fit Carter's popular image as a humble peanut farmer.

Recent presidents and first ladies have used the White House Grounds as settings for children's activities they seek to promote. George W. Bush loved baseball. He played Little League ball as a youth, and as president he invited local area T-ball teams to compete on the South Lawn. Bush's "White House Tee Ball Initiative" sought to encourage children's interest in baseball and softball and build the spirit of teamwork.

First Lady Michelle Obama's "Let's Move" program brought children to the South Grounds once again to play games. Encouraging children to play outside was just one aspect of her initiative to end childhood obesity, which also included the promotion of healthy eating and the establishment of a private-public foundation called Partnership for a Healthier America. "The physical and emotional health of an entire generation and the economic health and security of our nation is at stake," said the first lady in launching "Let's Move" on February 9, 2010. By playing and dancing with children, she showed that healthy activities can also be fun.

Presidential Recreation

MOST NINETEENTH-CENTURY PRESIDENTS did not add recreational facilities to the White House Grounds because of the near-constant presence of the general public. One notable exception was a croquet court installed in front of the South Portico by Rutherford B. Hayes in 1878. His family enjoyed the game and often played it together and with friends. Although the croquet court was not made available for public use, President Hayes allowed his White House clerks to enjoy a few games during business hours, considering them his "office family" deserving of breaks from their hard work.

President Theodore Roosevelt enjoyed sports, especially tennis, which he often played with the younger members of his staff, also known as his "Tennis Cabinet," seen here with the president in 1909.

The first putting green, installed for President Eisenhower, remained in place until the 1970s. In 1995 a new putting green (seen here c. 2013) was installed for President Clinton.

In the spring of 1903, Theodore Roosevelt ordered a regulation tennis court constructed on the near South Lawn just outside the West Wing. As with all courts of the era, the playing surface was hard-packed dirt. Owing to the dense packing of the underlying earth, it exhibited a great amount of spring and was considered to be one of the finest courts in the country. The entire Roosevelt family played tennis, and Alice was particularly noted for her skill. Newspapers observed that she took up the sport "with all the enthusiasm that a daughter of President Roosevelt might be expected to." Not surprisingly, the president himself gained a reputation for a "fast and aggressive style of play," and although the majority of games were played without keeping score, it was common knowledge that he won far more contests than he lost. At the time, tennis was still relatively new to the United States, and its subsequent national popularity was attributed to the presidential family's interest. The *New York Evening Post* even claimed that since "the most strenuous occupant that ever tenanted the White House" was an avid tennis player, it could no longer be easily dismissed as a "pastime for mollycoddies."

In 1954 a one-hole putting green was installed on the South Lawn for President Dwight D. Eisenhower, here seen practicing his swing near the South Portico.

On May 10, 1922, President and Mrs. Warren G. Harding held a tennis tournament for national and international players on the White House Grounds. Here the president and first lady watch a match from the sidelines.

The Roosevelt tennis court was removed in 1909 to make room for an expansion of the Executive Office under President William Howard Taft. A new court was subsequently installed further down the South Grounds near the location of the current swimming pool. In May 1922, President Warren G. Harding and First Lady Florence Harding invited a major international tournament to be played on this court, thereby further elevating the President's Grounds as a setting for serious tennis. Tragedy struck two years later when President Coolidge's 16-year-old son, Calvin, Jr., died of blood poisoning incurred from a blister he developed on his right foot while playing tennis on the White House court. Shortly thereafter, the tennis court moved once again, this time to a location on the far South Grounds due west of the large fountain basin. Most recently, President Barack Obama refitted the White House tennis court for double-duty as a full-size basketball court.

President Herbert Hoover liked to exercise with a medicine ball in order to keep in shape. He and the members of his cabinet also played "Hooverball," a volleyball-type game invented by the president's physician. Making use of a specially constructed court on the far South Grounds, two teams took turns throwing a 6-pound ball back and forth over an 8-foot net, scoring points whenever a player failed to catch the ball, return it back over the net, or caused it to land out of bounds.

Golf was President Dwight D. Eisenhower's favorite sport for exercise and stress relief. After he became president, security limited his use of public and private courses. Consequently, in February 1953, the American Public Golf Association struck upon the idea of donating a one-hole putting green and having it installed on the South Lawn of the White House. A year later the green was in place, but the grass failed to grow well and had to be replaced by a special variety known as "Pen-lu sod." This new grass worked much better, and National Park Service groundskeepers meticulously maintained the green. They replaced damaged portions of sod as needed from a dedicated nursery in nearby Maryland and every week moved the single hole so as to provide a variety of new challenges.

President Gerald R. Ford and his daughter Susan observe the work in progress as a hole is excavated for the installation of a swimming pool on the White House Grounds in May 1975. The president's first swim attracted a great deal of press attention.

Mature trees create a private setting for the White House swimming pool today.

The putting green harmonized so well with its surroundings that squirrels became something of a nuisance. Radio broadcasts and newspapers circulated unfounded reports that they were all going to be either killed or trapped and removed wholesale from the grounds. In reality, only three particularly troublesome squirrels were humanely trapped and released into nearby woodlands before letters expressing concern over the fate of the animals started appearing in increasing numbers in the heaps of mail delivered to the White House each day. Children sent their pocket money, dimes, and quarters, even dollars, to pay for a fence to keep the squirrels out of the putting green. Whenever possible the funds were returned to the senders, along with a formal thank-you from National Park Service Assistant Superintendent Frank Gartside. One letter from a would-be donor to President Eisenhower expressed the honest audacity only a child can muster: "Sure you are President of the United States. But God made the squirrels too and he put them there a long time before you ever thought of being President. . . . After all those squirrels have seniority over you." To be clear, no squirrels were ever harmed on account of the putting green. The letter-writing campaign proved successful, and "Operation Squirrel," as the limited trap-and-release program came to be known, was quickly suspended.

Gerald Ford, star player on the University of Michigan football team and one of the most athletic presidents in American history, had the present in-ground swimming pool and adjacent cabana built. This new construction cost the American people nothing, as it was financed by private donations. Through the study of numerous design proposals and existing conditions assessments, the planning team selected a location south of the West Wing because it afforded the necessary privacy and safety features while impacting the historic landscape to the least extent possible. During the pool's excavation, the government unearthed a substantial amount of archaeological materials, including burned bricks and a marble fragment that is part of an original White House mantel. These discoveries indicated that the site functioned as a trash pit following the 1814 fire and show that there is always something new to learn about the rich history of the grounds.

President Barack Obama shoots hoops on the White House basketball court (adapted from its usual use as a tennis court) during a break between meetings, 2010.

Pitching horseshoes was a favorite activity of President George H. W. Bush, and so he had a pit installed on the South Lawn in 1989.

South Korean President Kim Young-sam gestures while jogging with President Bill Clinton on the White House track in November 1993.

A horseshoe pitch was installed near the east side of the swimming pool in 1989 so that President George H. W. Bush could enjoy a favorite hobby while at the White House. His successor, President Bill Clinton, removed the horseshoe pit and installed a jogging track around the edge of the roadway on the South Grounds. George W. Bush, in turn, had the jogging track removed and the horseshoe pitch replaced. Eight years after leaving the White House, the elder Bush returned to pitch a game with his son, the new president of the United States.

THE WHITE HOUSE
WASHINGTON

First Children and Pets at Play

THE AVERAGE AGE of presidents upon taking office is nearly 55, and often their children are grown, so when presidents do bring young children to live at the White House they are always of special interest to the public.

President Abraham Lincoln's youngest son, Tad, was just 7 when he moved into the White House, and he immediately made the mansion and its grounds his own personal playground. Like most boys, Tad played war, firing off invisible bullets from make-believe guns, but unlike most boys he was surrounded day and night by an entire company of Union soldiers who quickly befriended him. Before long, the boy was taking a regular place in the chow line and drawing rations of his own.

"Third Lieutenant" Tad Lincoln had a well-known mischievous streak. Bestowed an honorary commission, he enjoyed joining and playing practical jokes on the soldiers of Company K who were encamped on the South Grounds to guard President Abraham Lincoln and his family during the Civil War.

Bo, President Barack Obama's Portuguese water dog, poses for the camera on the North Lawn in 2012.

Tad would not long remain in the ranks of the enlisted men, however. Before the close of 1862, he received an honorary officer's commission complete with the fictitious rank of "Third Lieutenant" and a child-size uniform. The new "officer" particularly enjoyed inspecting and drilling the men outside the White House. Often forming squads with his friends in the infantry guards, who invariably played along with good humor, he reviewed them from atop his pony, saber at the ready. On one occasion he determined to give the soldiers the night off by replacing the regular sentries with conscripted members of the White House domestic staff. According to the newspaper correspondent Noah Brooks, the president's eldest son, Robert, home from Harvard, protested that the servants should not be "compelled to do special duty when their day's work was done." Nonetheless, Tad staunchly "insisted on his rights as an officer," and President Lincoln, who found the whole situation extremely humorous, did not intervene until Tad fell asleep and then he went outside and "personally discharged the sentries his son had put on post."

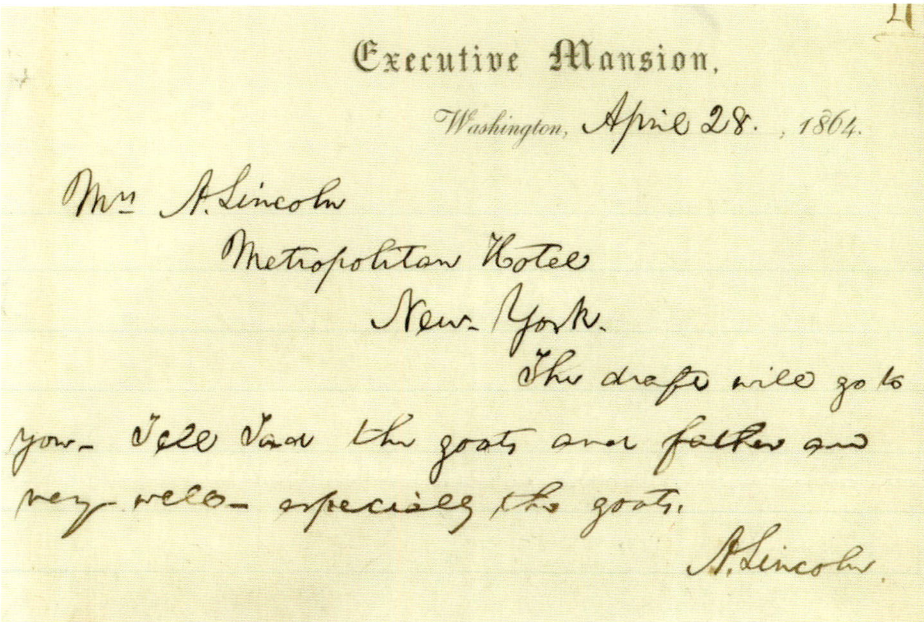

Executive Mansion,
Washington, April 28., 1864.

Mrs A. Lincoln
Metropolitan Hotel
New-York.

The draft will go to you. Tell Tad the goats and father are very well—especially the goats.

A. Lincoln.

The rambunctious Tad Lincoln, seen riding his pony on the White House Grounds in November 1863, was famously fond of his goats. In April 1864, President Lincoln wrote to his wife directing her to "Tell Tad the goats and father are very well—especially the goats."

Even when not playing out his fantasies as a Union soldier, more often than not Tad could be found playing outdoors on the President's Grounds. He could count on his soldier friends for a good story or an impromptu race, and there were plenty of good climbing trees. There were also dogs to play with and two pet goats, Nanko and Nanny. Many years after the war, former Union nurse Rebecca Pomroy fondly recalled that she often saw Tad happily "drawing his dogs tandem over the lawn." Late in life Colonel William H. Crook, who began his service at the White House as a plainclothes guard for Lincoln and stayed on to serve as disbursing officer for every president up through Woodrow Wilson, fondly recalled Tad's good-natured attitude whenever the goats refused to cooperate:

> As I look back over nearly half a century, I see him most plainly and oftenest seated in a little wagon, driving a pair of goats around the White House grounds. Sometimes the goats would trot along as directed, and sometimes they would decline to move, or move in the wrong direction, or try to move in two directions at once, as goats will. But Tad didn't mind much. He would simply wait until the steeds were ready to go where he wished, and then they would start on.

While vacationing with his mother in the spring of 1864, Tad began to worry about Nanko and Nanny and sent his father, who had remained at the White House, a telegram asking him to check on them. Lincoln's quick reply shows that no matter how busy he was or how urgent his responsibilities, he was always a loving and devoted father. It also reveals his dry humor. Addressing the telegram to Mrs. Lincoln, he wrote: "Tell Tad the goats and father are very well—especially the goats."

Forty years later, it was once again common to see a child with a goat and wagon out on the President's Grounds. Little Benjamin Harrison McKee, President Harrison's grandson who lived in the White House with his fam-

"His Whiskers," President Benjamin Harrison's mild-mannered pet goat, was enlisted to pull a cart to carry the president's grandson, Baby McKee, around the White House Grounds.

ily and was affectionately called "Baby McKee" in the press, loved being pulled along by his pet, aptly named "His Whiskers." One afternoon, when Baby McKee sat in a wagon hitched to the goat, the usually mild-mannered animal became alarmed and took off running down East Executive Avenue. The president sprinted after the pair, bringing them to a halt at the edge of an open excavation. The newspapers reported that "Baby McKee was as unconcerned as could be but his distinguished grandfather was pretty well winded as he led the goat and cart back to the White House."

The Harrison grandchildren were followed by the Cleveland children, and although the Clevelands had closed the gates to the President's Grounds, people routinely lined the fences hoping to see little Ruth Cleveland, known in the press as "Baby Ruth," at play. Reporters stationed themselves within the crowd to take notes on the child for stories that were picked up by newspapers across the nation. They reported on March 25, 1894, for example, that Ruth spent the day "tripping after the soft grass, dipping her nose into the opening spring blossoms like a gay little hummingbird," and "picking dandelions and blue bells on the southern lawn of the White House grounds." Farther off in the distance, a nurse wheeled her baby sister, Esther, around in a carriage underneath the trees.

Theodore Roosevelt once remarked, "I don't think any family has ever enjoyed the White House more than we have." When he took office in 1901, his large, active family included six children, all under age 18: Alice (17), Theodore Jr. (14), Kermit (12), Ethel (10), Archie (7), and Quentin (3). They had at least eight dogs at the White House. Pete the bull terrier was probably the most famous, or infamous, as he bit so many legs that he was eventually sent back to the family home in Oyster Bay, New York. Newspaper accounts report that the feisty little dog excelled in terrorizing high-ranking officials and diplomats. He reportedly chased French Ambassador Jean Jules Jusserand and George Cortelyou, who served in Roosevelt's cabinet, up trees on the President's Grounds, nearly sank his teeth into Secretary of the Interior James R. Garfield, and overran Assistant Secretary of State John O'Laughlin on numerous occasions. Reporters described Pete as the White

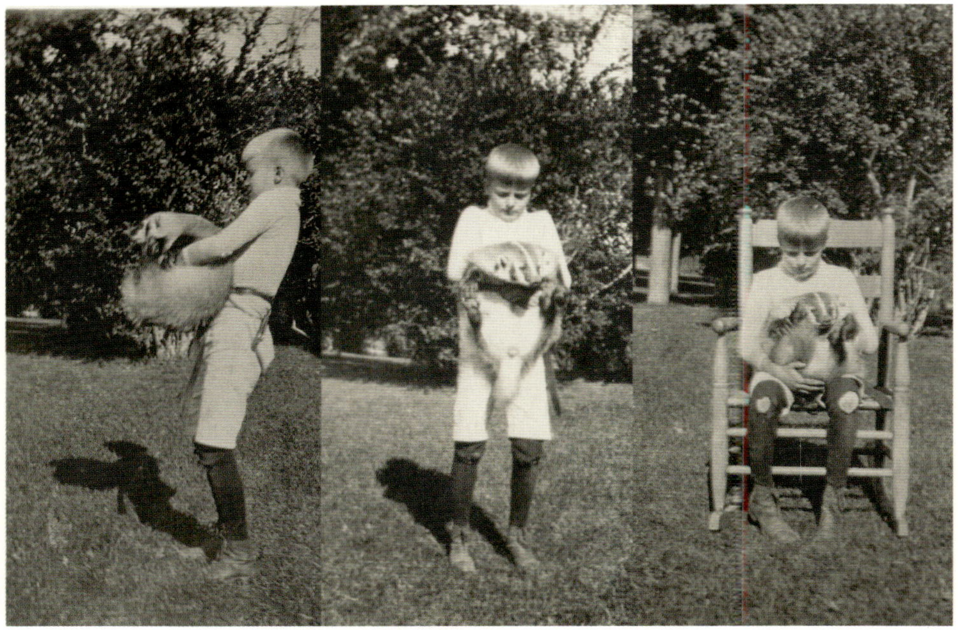

President Roosevelt's family brought at least eight dogs to the White House including Rollo, a Saint Bernard, seen with the president (above left), and Skip, a mutt that befriended the president during a hunting trip, seen with the family (opposite). The dogs had free rein over the grounds, and the children loved playing with them. Their more unconventional pets included a one-legged rooster and a badger named Josiah, pictured with Archie (above right).

House watchdog, always patrolling the grounds for "an appetizing statesman or a juicy clerk." Among the latter was John T. Thomas, who lost a tug-of-war over his trousers. After the White House policemen finally pried Pete loose, a visibly shaken Thomas was quoted as saying: "I have been told it is a distinction to be bitten by the President's dog, but I would prefer not to be the centre of so much interest."

Algonquin, Archie's calico pony, is best remembered for playing the lead role in a famous stunt while the child was confined to bed with the measles. Hoping to speed his brother's recovery, Quentin, the youngest Roosevelt, stealthily brought the pony in through the basement and up the elevator to Archie's Second Floor bedroom, much to the sick boy's delight and the bewilderment of the domestic staff and resident police force. This indoor adventure often overshadows the stories of Skip, a much-loved mutt of a dog, who would jump on Algonquin's back and go for rides around the grounds while he barked with delight. The two animals made for an unusual sight, and they were clearly the best of friends.

The Roosevelts had a menagerie of other pets as well. On any given day the children frolicking on the South Lawn might be joined by cats, guinea pigs, snakes, rats, squirrels, pigs, goats, cows, horses, and even an opossum. A blue macaw, Eli Yale, lived in one of the greenhouses and ate bread, coffee grains, and potatoes. Josiah the badger, whom Roosevelt acquired as a baby while campaigning in Kansas, also ate potatoes and loved milk. He played tag with the children on the White House Grounds and occasionally nipped at their bare heels. Archie would often pick Josiah up and carry him around by the waist. Even the fearless President Roosevelt thought this a bad idea, but Archie admonished his father's "unworthy assault on the character of Josiah." "He bites legs sometimes," said Archie, "but he never bites faces."

The children kept a horned toad named Bill in a box on the South Portico for many years, and Kermit housed a one-legged bantam rooster in a coop underneath the steps. Roosevelt forbid his boys from entering the bird in fights, but the press made a not-so-subtle pun in declaring that the rooster looked "capable of winning a championship in the featherweight class." The

Like the Roosevelts sixty years earlier, the young Kennedy family enjoyed ponies and dogs at the White House, and even took them on vacations. The first family is seen here with their dogs Clipper, a German shepherd; Wolf, an Irish wolfhound; Charlie, a Welsh Terrier; and puppies. Caroline was especially fond of Macaroni and she often rode him around the White House Grounds. She is seen here with Macaroni and her brother John visiting her father outside the Oval Office in 1962.

Young Quentin Roosevelt rides the calico pony, Algonquin, on the South Lawn, c. 1902. President Theodore Roosevelt made doodles of his children's escapades, including the sketch above of Ethel taking a tumble off her pony.

rooster escaped on occasion and despite his disability usually took more than an hour to corral. There was even a young black bear, but he bore little resemblance to the famous Teddy bears flying off store shelves. The president explained that the bear's name, Jonathan Edwards, was in part in homage to Edith Roosevelt, whose family descended from the Puritan minister, and "partly because the bear possessed a temper in which gloom and strength were combined in what the children regarded as Calvinistic proportions."

The Kennedys arrived in the White House with very young children, and Vice President Lyndon Johnson gave them two ponies, Macaroni and Tex. Both ponies roamed freely on the President's Grounds but otherwise lived in a renovated maintenance building that is still known as the "pony shed." The Kennedys brought a variety of other pets with them to the White House, including dogs, rabbits, sheep, and guinea pigs, all of which were kept in the children's playground outside the West Wing. The playground itself featured a tree house, swing, slide, barrel tunnel, and even a trampoline, all built according to First Lady Jacqueline Kennedy's specifications. Mrs. Kennedy even enjoyed bouncing on the trampoline, and she cherished time outdoors with her children.

Amy Carter was just 9 years old when the Carters moved into the White House. She played in a tree house on the grounds designed by her father, President Jimmy Carter. It was built as a self-supporting platform so as to not to damage the Blue Atlas cedar tree it rested on.

The Obamas, too, brought young children to the White House—Malia, age 10, and Sasha, age 7. Throughout their time in the White House the Obamas advocated for outdoor play and exercise as an important component of a healthy lifestyle and physical fitness plan.

President Carter, his 9-year-old daughter Amy, and 1-year-old grandson Jason Carter, enjoy the tree house designed by the president on the South Grounds, c. 1977.

In 1933, President Franklin Roosevelt installed playground equipment, swings, and a sandbox on the White House Grounds for his grandchildren.

Shortly after taking office in 2009, President Barack Obama and First Lady Michelle Obama had a small playground built for their daughters on the west side of the South Grounds within view of the Oval Office. The equipment included a swing set, rope ladders, slide, and crow's nest. This playground was a surprise, installed when the girls were at school. Upon returning home, they played for nearly an hour, despite the fact that it was a cold day with snow still on the ground. The president is seen here with Malia and Sasha as they enjoy the tire swing.

THE WHITE HOUSE GROUNDS DURING WARTIME

DURING WARTIME, civil liberties are often temporarily restricted, and so it has been with access to the President's Grounds. A state of war changes everything for the president, and for the nation, and the White House Grounds have always been specially guarded during wartime and sometimes even turned over to military purposes. The following account covers the War of 1812 and, especially, the Civil War, as well as the two world wars of the twentieth century. Most recently, however, it has been terrorist attacks that have tightened security and closed the grounds to visitors. Following the 1983 bombing of the U.S. Embassy and Marine barracks in Beirut, Lebanon, concrete barriers were placed outside the White House perimeter. Then, in 1995, after the bombing of the Alfred P. Murrah Federal Building in Oklahoma City, the streets surrounding the White House were closed. Tours of the White House continued until the terrorist attacks of September 11, 2001. Today visits must be arranged through a member of Congress, and visitors undergo a security check in advance. But twice a year—in the spring and the fall—the President's Grounds are open to visitors through a timed ticketed system.

POLICE
POLICE LINE DO

The War of 1812

WHEN WAR FIRST IMPACTED the White House Grounds, it did so in the worst way possible. On June 18, 1812, during the administration of President James Madison, the United States declared war on Great Britain, following years of violations of American free trade and freedom of the seas. The Americans fared poorly in the first years of the war, and by the summer of 1814 the British ruled the Chesapeake and were planning a land invasion. In response, a company of one hundred U.S. soldiers set up camp on the North Lawn and mounted cannon at the North Gate. Most of these men were, however, subsequently ordered into the field, leaving the city virtually undefended.

This 1814–15 watercolor, The President's House *by George Munger, depicts the aftermath of the fire of 1814. Interestingly, the painting seems to show a largely barren North Lawn, raising the possibility that the British may not have spared the grounds, as is commonly believed. One theory holds that the saplings in the foreground are young ailanthus trees, also known as Tree of Heaven, a fast-growing species that thrives in the wake of fire.*

After the debacle at Bladensburg, Maryland, on August 24, where the British routed the American forces only a mile outside Washington, President Madison's chief steward, Jean-Pierre Sioussat, recommended spiking the cannon and leading a trail of gunpowder to the front door, ready to destroy the mansion rather than see it fall to the enemy. First Lady Dolley Madison forbade this extreme action, but like everyone else, she was helpless to prevent the British from setting fire to the White House, Capitol, and executive offices that night. Only hours before, she famously ordered Gilbert Stuart's full-length portrait of George Washington removed for safekeeping. Paul Jennings, one of the Madisons' slaves, recalled much later that Sioussat and Thomas McGrath, the first person known to have served as gardener at the White House, took down the painting from its place of honor in the State Dining Room and "sent it off on a wagon, with some large silver urns and other such valuables as could be hastily got hold of." Only then did the first lady flee, joining her husband who had been traveling with the army for several days.

The Madisons returned three days later to find only the soot-stained outer walls of the White House still upright. Thomas Jefferson's wall of enclosure survived, as did the monumental entrance arch at its southeast corner, but the state of the landscape immediately following the fire is largely unknown. The White House historian William Seale contends, "It had been the neatest of fires. The refuse had fallen precisely within the stone walls, and no debris was scattered over the outer grounds." This reasoning coincides with period accounts that describe the burning as a professional, even ceremonial, military exercise. In burning the public buildings, the British dealt a serious blow to the American psyche, and ruining the grounds was unnecessary. Beyond the inevitable collateral damage, they probably left the landscape alone.

If the fire did not substantially damage the President's Grounds, the reconstruction of the mansion certainly did. Under the supervision of James Hoban, the original architect of the White House, the rebuilding effort began with tearing down those portions of the structure that were not salvageable. In need of a place to dump this debris, the workmen used a gully southwest of the mansion. Work progressed steadily, and by February 1817 a new roof was supported by restored stone walls. During these busy years much of the work already accomplished on the landscape seems to have been disrupted or destroyed. Although not completely ruined, the President's Grounds essentially reverted back to the construction site it had been less than twenty years prior, during the initial construction of the White House, complete with temporary buildings and large piles of buildings materials and garbage.

Consequently, the site was soon in need of its own redevelopment, and the now-missing 1818 landscape plan by architect Charles Bulfinch must have been the model, itself likely based on Jefferson's concept and vision.

THROUGHOUT THE CIVIL WAR, the White House symbolized the Union cause. President Abraham Lincoln used it to promote patriotism and support the war effort. The outdoor Marine Band concerts, for example, incorporated patriotic songs to rally public support, and they were never once canceled throughout the war, although temporarily moved to Lafayette Park after the death of young Willie Lincoln in 1862. As president, Lincoln sought to project an aura of strength and confidence, even when the actual fighting was not going well.

Lincoln dedicated a flagpole on the South Grounds on June 29, 1861, with a military ceremony that included the salutatory firing of two pieces of field artillery. This event was undoubtedly conceived as a show of solidarity during the first, uncertain, summer of the Civil War. The distinguished guests included Lincoln's entire cabinet, the mayor of Washington, eleven senators, and Congressman Schuyler Colfax, future Speaker of the House of Representatives. Each received a handwritten invitation requesting his presence at the gathering, officially described as the "erection and dedication of a staff and Flag of the Union."

This painting by Albert Bierstadt depicts Abraham Lincoln's flagpole on the South Lawn, with the Potomac River in the background, c. 1863.

Artist Alfred R. Waud captured the military ceremony in which Lincoln's flagpole was dedicated on the South Grounds in this sketch titled Flag Raising at the White House, June 29, 1861.

Almost three years later to the day, on July 4, 1864, President Lincoln hosted a larger and much different group on the South Lawn when he allowed a congregation of African Americans to assemble and stage "a grand Sunday-school festival." Fittingly, the artist Francis Bicknell Carpenter, who was just finishing his famous painting, *First Reading of the Emancipation Proclamation by President Lincoln*, attended the event. Before his general Emancipation Proclamation went into effect on January 1, 1863, freeing all the slaves in Confederate-held areas, such a meeting outside the White House would have been all but impossible. Until this time, Lincoln had maintained that the only objective of the war was the restoration of the Union, and presiding over such a large number of African Americans on the President's Grounds could easily have been interpreted as a pro-abolitionist stance. After the proclamation, however, Lincoln never wavered in his conviction that winning the war now also entailed the end of slavery in the United States. Hosting the African American Sunday school called public

a sketch of the white house is in our paper

attention to this change in policy, and doing so on the Fourth of July—Independence Day—made the message all the more clear.

Carpenter was pleased to report that no day ever "presented a busier or more jubilant scene" or a "greater appearance of enjoyment and success." He described how the audience listened intently to a series of speakers who stood on a specially constructed platform. Later, children and adults made good use of swings hung from the sturdy boughs of strong trees, relaxed on the soft grass, and strolled along the shaded paths of the White House Grounds. The highlight of the event came when a committee of two laymen and three clergymen solemnly presented President Lincoln with an elegant copy of the Holy Bible. Its front cover depicted Lincoln as the "Great Emancipator," standing in a cotton field and "knocking the shackles off the wrists of a slave." The back cover featured an inscription praising the president as "the friend of Universal Freedom."

Not everyone in the national capital was happy about this display of racial equality. The editors of the *Constitutional Union*, a local newspaper that was no supporter of African American rights, described the religious gathering as a picnic: a "vast herd of negroes" crowded together "beneath the very eaves of the building erected by white men for the residence of the Chief Magistrate of white men," and Lincoln's emancipation of the slaves "the bitterest cup of woe ever held to a nation's lips." That the article was reprinted in many northern newspapers is an indication of a widespread opposition to the notion of racial equality, if not to slave emancipation.

President Lincoln proved immune to this criticism, and he allowed a second African American Sunday school to occupy the White House Grounds one month later. This event culminated with the president receiving a "Banner of Freedom." According to Reverend Jean R. Pierre, the event's principal organizer, this banner featured a life-size likeness of Lincoln allegorically leading a group of slaves out of plantations and into a church. Again the *Constitutional Union* defamed the event as a "drunken revelry."

For much of the Civil War, the White House and its grounds were protected by a contingent of Union soldiers. Even before Lincoln arrived in Washington for his inauguration on March 4, 1861, Colonel Charles P. Stone, commanding officer of the District of Columbia Militia, had mobilized his forces to secure the city. Lincoln's swearing in took place without incident, but Colonel Stone, an ardent nationalist and the first volunteer officer in the Union Army, made certain that the new president's first night in the White House was peaceful too, for he stationed his men in concealed positions outdoors behind trees and shrubbery.

A company of some 125 additional soldiers under the command of Kansas Senator James H. Lane joined the militia during the night of April 19. Following a midnight review by Lincoln on the President's Grounds, sixty members of this self-described "Frontier Guard" retired to the grand East Room, where they "slept upon their arms." Senator Lane, an active abolitionist, raised the company from like-minded Jayhawkers who had traveled to Washington in order to develop contacts within the Lincoln administration. The president probably did not expect that his request would be taken so literally, especially not in the middle of the night, but he raised no objection, and Lane was certainly not the only well-motivated Union man to expect a personal reception with the commander-in-chief. Local papers initially reported that the Frontier Guard would be on patrol outside the White House each night beginning at 8:30 p.m., but Lincoln, whose personal sensibilities were already strained by the continued presence of Stone's troops on the grounds, quashed the notion. He consented only that Lane's men be located close enough to "rally there at the first alarm."

Only a day before Senator Lane's arrival, fellow abolitionist Cassius M. Clay raised an emergency force of Union loyalists from among the men, both locals and visitors who happened to be in Washington at the time. In his autobiography written twenty-five years later, Clay recalled that the improvised group included ex-congressmen, state governors, and "other men of mark," but he made only a vague statement that his force aided in "guarding the President's house," so the full extent of its involvement is not known. His presence would not be surprising, as he greatly valued their volunteerism. When federal troops arrived in Washington at the end of April, they relieved the citizen-soldiers under Lane and Clay, but not

President Lincoln and General Winfield Scott review Union volunteers from the North Lawn in this pencil sketch by Alfred Waud, July 1861.

before Lincoln sent a special dispatch to the latter extending his praise and "high appreciation of the very prompt and patriotic manner in which your Battalion was organized for the defense of the capital, and the very efficient services rendered by it."

The federal troops continued pouring into the city throughout May and June 1861, and they often drilled on a parade ground south of the White House on what is now the Ellipse. President Lincoln repeatedly came out to view their maneuvers, and other spectators watched from atop Jefferson's old ha-ha wall. As impressive as these military exercises were, they paled in comparison to the spectacle that unfolded along the North Front of the White House on July 4, 1861, when thousands of Union Soldiers marched down Pennsylvania Avenue. A high wooden reviewing stand had been erected outside the mansion for the occasion. Built up against the North Fence, this stand featured a brightly colored canvas canopy that was in fact an immense American flag. A second, smaller flag, attached to a pole at the front of the stand, flapped in the wind. Shortly before the soldiers began their march, the president crossed the North Lawn and ascended a flight of steep wooden stairs. Once on top of the platform, he elected to remain standing as he reviewed the troops passing by. The parade went on for hours, with the effects of the hot summer sun only partially alleviated by the shade canvas.

Later that month, the Union Army departed Washington for Manassas, Virginia, where the first major engagement of the Civil War occurred. Colonel Stone and his men marched out with the rest of the soldiers, leaving only the Washington Metropolitan Police to keep watch over the White House. Soldiers did not return to guard the mansion for more than a year, when a show of force by the Confederates in September 1862 raised concerns for the security of the national capital and the safety of President Lincoln. General Robert E. Lee's advance into neighboring Maryland set off a full-scale panic in Washington. To prevent an invasion, Union troops streamed back into the capital. Among them was the 150th Pennsylvania Volunteer Infantry Regiment, nicknamed the "Bucktails" for the deer tails that adorned their caps. The regiment's Company K soon received orders

A guard stands at his post at the Northeast Gate to the North Portico of the White House in this 1861 photograph attributed to Montgomery C. Meigs. The gates were removed in 1870.

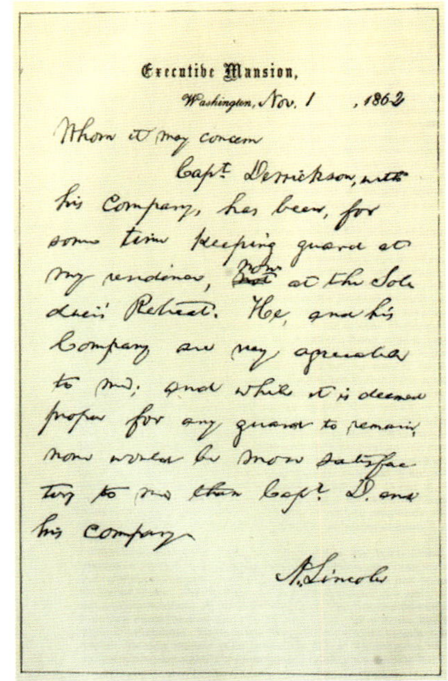

Executive Mansion,
Washington, Nov. 1, 1862.

Whom it may concern

Capt. Derrickson, with his Company, has been, for some time keeping guard at my residence, now at the Soldiers' Retreat. He, and his Company are very agreeable to me, and while it is deemed proper for any guard to remain, none would be more satisfactory to me than Capt. D. and his Company.

A. Lincoln

Approximately sixty-five men of Company K served as the Lincoln family's personal guard throughout the Civil War and operated out of a camp on the White House Grounds seen here above Jefferson's ha-ha wall. On November 1, 1862, Lincoln issued orders for Company K to remain on the grounds. For most of this time the men lived in tents, until they moved into a wooden barracks in January 1865.

to guard President Lincoln and his family and in this capacity established a camp on the South Grounds of the White House in November 1862.

All told, approximately sixty-five Bucktails bivouacked beneath the windows of the mansion for fully half the Civil War. Lieutenant Colonel Thomas Chamberlin, one of the highest ranking officers to serve in the 150th Pennsylvania and the author of its regimental history, recorded that the company "pitched its camp on the lawn a short distance southeast of the mansion. Here it remained during the war." Most of the information about the soldiers' camp on the White House Grounds comes from the personal papers of William Cutter, a private in Company K. From the fall of 1861 to the early spring of 1865, he wrote some 150 letters to his recently widowed mother, brothers, sisters, and other family back home in Meadville, Pennsylvania. His faithful correspondence provides details about daily life in the camp, guard duty, and his personal thoughts and feelings toward the Lincolns.

Though nearly winter, Cutter's first night on the South Grounds was mild, and a welcome surprise for the men of Company K who were not furnished with winter quarters and did not even have a chance to set up their tents until the following afternoon. Cutter improvised "a good bed" from his rubber poncho, wool blanket, and overcoat, as he wrote his mother. Day and night a pair of Bucktails patrolled the North Grounds, while others stood watch at the Pennsylvania Avenue gates and guarded the company camp. Cutter marveled that after the sun went down the gasoliers on the carriage gates burned so brightly that he could still "see to pick up a pin it is so bright."

The following February, the 150th Pennsylvania received orders to depart Washington and head into the field as part of the Army of the Potomac. Inexplicably, this action initially included Company K, and by mid-month the men were all packed and ready to move out. Then, at the last possible moment, First Lady Mary Todd Lincoln intervened and begged them to stay. In seeing the soldiers making ready to depart, the president had declared that if Company K would not be his guard, then "he would have no guard," and she feared that he would make good on this promise. Thus caught between his orders and the first lady's request, which clearly reflected Lincoln's own opinion, the company's lieutenant, then in temporary command, took the unprecedented and borderline mutinous step of putting the matter to a vote. The men hastily assembled on the South Lawn and, except for one unnamed soldier, all elected to stay. They drew up a short note testifying to this decision and, after Lincoln added his signature, Company K became officially detached from the rest of the regiment. Upon hearing the news, an outraged Colonel Langhorne Wister, commanding officer of the 150th Pennsylvania, stormed over to the White House to protest, but Lincoln refused to see him. The frustrated colonel led his remaining nine companies out of Washington the very same day. Cutter wrote home to his mother, "We are left behind to guard the president as long as he wants a guard."

Despite the lopsided vote, the decision to stay was not an easy one. From the minute they first arrived at the capital, the members of Company K had been itching to prove themselves on the front lines. In the coming years, particularly after their comrades in the 150th Pennsylvania served with distinction at Chancellorsville and Gettysburg, several men would regret their choice and seek transfers in pursuit of their own glory. Cutter probably summed up the sentiment of his entire company when he informed his mother in the spring of 1864 that, "For my part I would like to have a little fright before I go home." Yet the soldiers also recognized that they were providing a critical and very prestigious service. As Cutter put it, "Guarding the president has got to be a big thing." Styling themselves the "President's Guard," the men of Company K took great pride in their role as the Lincolns' personal protectors. Cutter wrote to his brother, George: "All the boys would of liked to go with the reg., but we could not leave old Abe when he wanted us so bad. We had to be his guard. Mrs. Lincoln said we should not go. She wanted Com. K to stay here. Tad Lincoln says he . . . was not a going to let us go, but we come darn near."

It is clear that a mutual affection quickly developed between the guards and the president's family. In particular, Lincoln counted the company commander, Captain David Derickson, as a friend. The two frequently breakfasted together, and Derickson became one of his most trusted advisers. On Christmas Day 1862, the president and first lady held a brief ceremony with the men on the South Lawn. Lincoln made a short, extemporaneous address in which he reportedly said that he considered the members of the company "a part of his family, and more than this, that there had never been any family jars." Cutter added that Mrs. Lincoln gave all the guards Christmas presents and served "four big gobblers" on New Year's Day 1863, the main course at a turkey dinner she had had prepared as a special treat. A year later, the company enjoyed an oyster supper as they celebrated the start of 1864 and what they hoped would be the last year of the war.

Cutter made sure to let his mother know about the food since she, like most soldiers' parents, was always concerned about his health and well-being. He actually gained nearly 20 pounds and confided to his brothers that he was "getting fat." As Cutter was finding out, serving as the president's guard clearly had advantages beyond being taken off the front lines.

Company K was not only well fed but also had ample time for recreation. Cutter's letters tell of a puppy that became something of a company mascot and a band formed by his fellow soldiers that boasted fiddles, banjos, tambourines, drums, horns, and various other brass instruments. Playing baseball on the parade ground immediately south of their camp at the White House was also a favorite activity, and Cutter, who was a frequent participant, seemed to especially enjoy the games.

There was also a swing, which was evidently hung from one of the large trees on the President's Grounds. Cutter wrote to one of his brothers on February 8, 1864, that a fellow Bucktail had been seriously injured in a bad fall: "One of our boys just now got hurt by falling while jumping off of the swing. He slipped and fell striking on his neck and shoulders." The young man was taken to a military hospital, and this seems to have been the only significant injury Company K suffered during the entire course of the war. The president's guards came under enemy fire only once, when they took part in the Battle of Fort Stevens. Fought July 11–12, 1864, the battle saw an assortment of regular Union troops, members of the Veteran Reserve Corps,

To mark the successful end of the Civil War, the Union Armies, under the commands of George Meade and William T. Sherman, paraded in front of the White House for two days. This color lithograph by Casimir Bohn, Grand Review at Washington: The Glorious Army of the Republic Passing the Head Stand, *captured the scene on May 23, 1865. The columns of the North Portico are wrapped in black in mourning for President Lincoln.*

and hastily armed clerks from the Treasury, War, and other government offices hold off a Confederate raid on the capital until federal reinforcements arrived. Dubbed "The Battle That Saved Washington," the only battle fought within the borders of the District of Columbia, the engagement was also notable by virtue of the fact that Lincoln rode out to view the fighting and became the only sitting U.S. president to come under enemy fire during wartime. When the smoke cleared Lincoln emerged unharmed, as did the Bucktails.

On November 8, election day, 1864, the men of Company K were able to vote in an impromptu polling station set up on the South Lawn, and they demonstrated their loyalty to President Lincoln by casting their votes, to a man, for him. "Comm. K went all right this time," Cutter wrote to his brother; "67 Union to one Copperhead and that one not by us boys." The "Copperhead" was another soldier who had been allowed to vote in Company K's camp and who cast his ballot for the president's Democratic challenger, George McClellan, formerly one of Lincoln's generals. Anticipating a Lincoln victory, Cutter added, "The Copperheads will find out in a few days where their little Mc is, for he hasn't a gunboat to get in to save his bacon as he used to have while in command here."

Private Cutter and the rest of the men of Company K left Washington to return home to Pennsylvania on June 15, 1865. It was just over two months since John Wilkes Booth assassinated President Lincoln at Ford's Theatre, and during the interim they had continued to guard the new president, Andrew Johnson. With the Civil War essentially over by the time of the assassination, the soldiers had not been sent along to guard Lincoln that fateful night. Had they been present, the plot to kill the president would likely have failed and American history forever altered. As it happened, the entire Lincoln family was devastated by the loss, including the soldiers of Company K.

World War I

THE WHITE HOUSE GROUNDS were closed to the public during World War I. The landscape was quiet, except for two things: the bleating of sheep and the rumbling of a tank. At its peak, a flock of eighteen sheep roamed over the landscape by day and bedded down for the night inside a 24 by 10 foot stable erected on the South Lawn. The sheep encouraged Americans on the home front to make the small sacrifices on behalf of the great sacrifices that U.S. soldiers were making abroad. To save on gasoline and manpower, the sheep took the place of lawn mowers, and if anything did too good a job. In addition to grazing on the grass, they clipped hedges and went after the flowers and other ornamental plants. Head Gardener Charles Henlock considered the flock more of a nuisance than a help, but the sheep became a popular symbol of wartime frugality. After sheering, they also produced enough wool to sell 2 pounds at auction to each state. At $10,000, Kansas paid the most for its allotment, and the entire fund-raising program totaled $52,823, all of which was donated to the Red Cross.

Sheep graze on the White House South Lawn near their water trough, c. 1919.

President Woodrow Wilson poses inside the official Liberty Bond tank during a demonstration on the White House Grounds on April 19, 1918.

If the sight of sheep on the White House Grounds was strange, then a tank was even stranger, for the iron behemoth was an entirely new war machine. In fact it was so new that the United States had not yet produced its own version of the weapon, and the average American citizen had never seen one. This particular tank was on loan from the British, who had developed the first combat model in 1916. It was sent to the United States as a publicity tool for the sale of Liberty Bonds, and it was in this capacity that President Wilson climbed aboard for an "impromptu" demonstration on the North Grounds on April 19, 1918. Upon exiting the vehicle, he burned his hand severely on the exhaust pipe but nonetheless beamed enthusiastically toward the vast throng of people assembled outside the front gates.

MILLER

World War II

WORLD WAR II was the last war to significantly impact the President's Grounds in the twentieth century. Although the landscape had been permanently closed to the general public for nearly two decades, security became such a concern that the military advocated repainting the White House in a camouflage pattern. President Franklin Roosevelt turned down this proposal but did allow a pair of anti-aircraft guns to be mounted on top of the mounds on the South Lawn.

Tightened security prompted moving the National Christmas Tree onto the White House Grounds for the 1941 holiday season. Since 1923, when President Calvin Coolidge presided over the lighting of the first National Christmas Tree, the tree had been located on the Ellipse or in Lafayette Park.

On Christmas Eve 1941, President Franklin D. Roosevelt and Prime Minister Winston Churchill lit the National Christmas Tree on the South Grounds.

A crowd gathered on the South Lawn of the White House on May 8, 2015, to observe the Arsenal of Democracy flyover of World War II vintage aircrafts. The event marked the seventieth anniversary of the end of the war in Europe on May 8, 1945, and commemorated the Allied victory.

In preparation for the 1941 lighting, two large Oriental spruce trees planted elsewhere on the South Grounds in 1925 were relocated to either side of the South Fountain. As the holidays approached, the eastern tree was decorated and strung with lights. On Christmas Eve, Roosevelt came out onto the South Portico to begin the ceremony. At his side stood British Prime Minister Winston Churchill, who had recently arrived for the first of many wartime meetings. The president pressed a button that signaled a workman to flip the switches, and an instant later the tree was aglow with light, much to the consternation of the Secret Service, who felt that the display made the White House an obvious target for an air raid, especially as the rest of the city was under a nighttime blackout. A small group of specially invited, security-cleared guests watched from inside the fence, while an estimated 15,000 people stood just outside the enclosure. Roosevelt and Churchill each made short remarks, but not before the spectators burst into song. So began the first Christmas of the war.

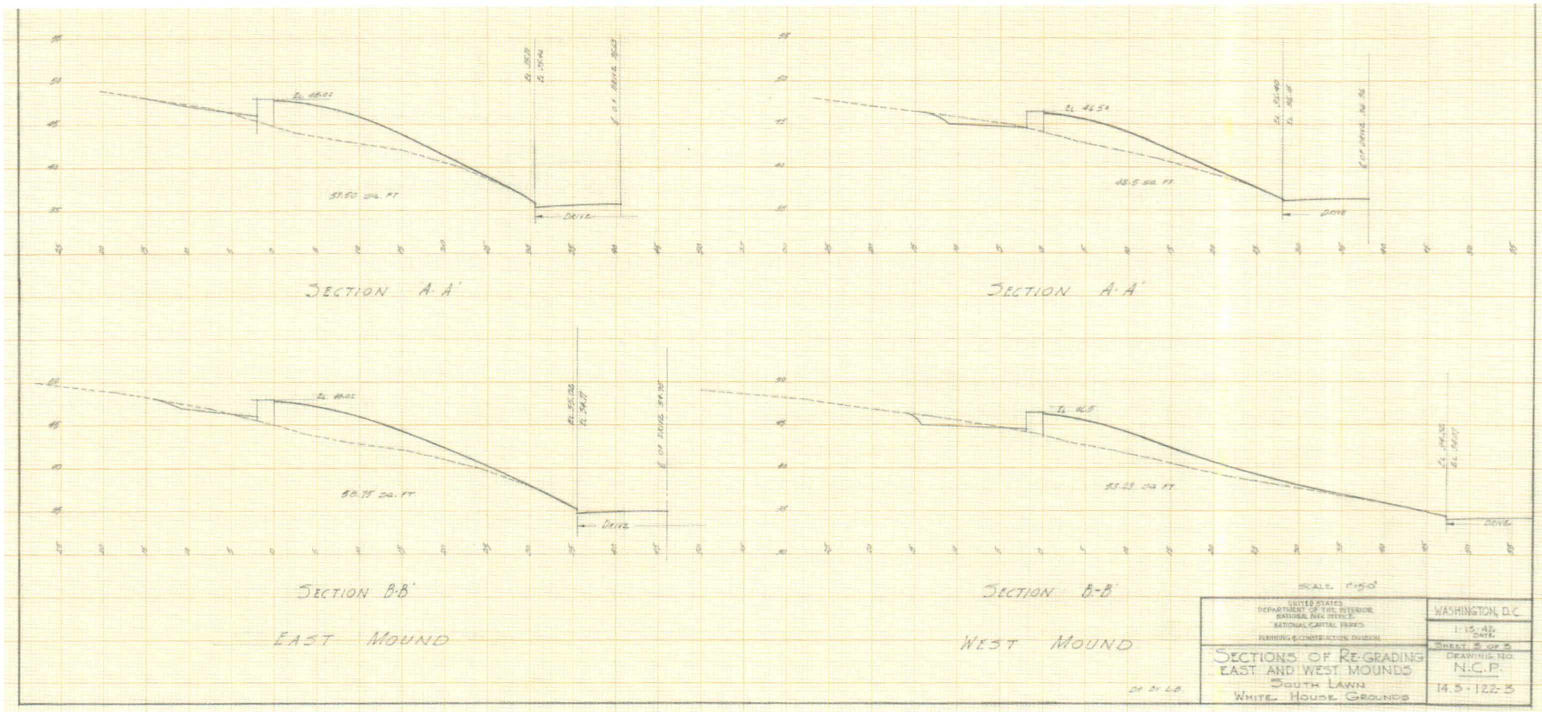

The slight changes in grade shown in this National Park Service drawing, dated January 13, 1942, were implemented in order to place anti-aircraft guns atop the pair of earthen mounds on the South Lawn. As a result, the mounds were flattened slightly and are now presumably slightly shorter than when they were first created in the nineteenth century.

Diana Hopkins, with her parents, tends to her Victory Garden on the South Lawn of the White House, May 11, 1943.

A newspaper seller in Lafayette Park, opposite the White House, hawks papers announcing unofficial reports of Japan's acceptance of Allied surrender terms on August 14, 1945. A crowd begins to gather on the grass to await official news.

The next year the scene was much the same, only Churchill was not in attendance and the western tree was decorated instead. In fact, the ceremony outlasted the war itself, and every year until 1953 one of the two spruces served as the National Christmas Tree. In 1954 the National Christmas Tree returned to the Ellipse, and today only the eastern tree is still standing on the President's Grounds.

In 1943, at the urging of First Lady Eleanor Roosevelt, one of the flowerbeds in the grassy lawn south of the East Corridor was taken up and the space converted into a demonstration victory garden. Tended by 10-year-old Diana Hopkins, the daughter of presidential adviser and former Secretary of Commerce Harry Hopkins, the little plot produced beans, carrots, tomatoes, and cabbages, but its real objective was to encourage Americans to follow suit in their own backyards and thereby allow more food to be sent to the soldiers serving overseas instead of grocery store shelves. This teaching tool, the first productive garden located on the White House Grounds in nearly three-quarters of a century, reverted back to lawn the following year, and the story was forgotten until 2009 when First Lady Michelle Obama planted a new vegetable garden on the South Grounds.

TOKYO RADIO SAYS
JAPAN SURRENDERS
TOKYO RADIO SAYS
JAPAN SURRENDERS
TOKYO RADIO SAYS
JAPAN SURRENDERS

PART III

WHAT GROWS ON THE PRESIDENT'S GROUNDS

THE TREES AS LIVING WITNESSES *192* / FOR THE TABLE *244*

ORNAMENTAL GARDENING *256*

THE TREES AS LIVING WITNESSES

WITH ALL DUE DEFERENCE to the iconic blooms of the Rose Garden, any discussion of the President's Grounds must begin with the trees. Many hundreds of grand and historic trees set off everything else that grows to great advantage. In the words of longtime Superintendent of Grounds Irvin Williams, "Trees tie it all down."

A History of Tree Planting at the White House

TRADITION ONCE HELD THAT Thomas Jefferson planted the first trees on the President's Grounds in the early 1800s, but there is no solid evidence to support this belief. A true Renaissance man whose many pursuits included landscape design and gardening, Jefferson continuously improved and redeveloped the grounds at Monticello and personally tended to plants indoors and out for more than four decades. He received many specimen plants from the finest nurserymen of the day, including Philadelphia-based Bernard McMahon, one of the fathers of gardening in the United States and something of a mentor to him.

Opposite and overleaf: On March 31, 1809, Nicholas King provided President James Madison with a planting list of "Trees and Shrubs for the President's Garden." The large trees listed are tulip tree, willow-leaved oak, elm, beech, ash, horse chestnut, English walnut, sugar maple, English sycamore, larch, weeping willow, mulberry (English, common, and Chinese varieties), and catalpa. The smaller trees include white pine, hemlock, mountain ash, common and honey locust, spruce pine, buckeye, redbud, Chinese arbor vita, golden willow, filbert, and holly. And the flowering trees and shrubs include laburnum, bladder senna, althea (white and red, double and striped), lilac (white, purple, and Persian), Scotch and Spanish broom, flowering almond and peach, evergreen thorn, privet, juniper, scarlet locust, yellow and white jasmine, honeysuckle (early, monthly and trumpet), bignonia, laurel, and rose.

Although Jefferson may not have planted at the White House, he did approve a planting plan in the waning days of his administration. Architect Benjamin Henry Latrobe, who supervised most of the building accomplished at the White House and on its grounds during Jefferson's presidency, informed James Madison about this plan on March 14, 1809, ten days after he became the fourth president of the United States. Unfortunately, this plan is now lost, but an extensive plant list dated two weeks later helps fill in many of the details. Written by City Surveyor Nicholas King and titled "Trees and Shrubs for the President's Garden," it lists fifty-one plants organized into three groups: large trees for single plantings or in lines, smaller trees for close plantings in clumps and screens, and flowering trees and shrubs. This classification is in keeping with the general landscape plan for the White House Grounds developed by Jefferson and Latrobe, which called for radial lines of trees on the North Grounds, vertical lines bordering an open lawn on the South Grounds, and a large area to the southeast marked "clump." It further indicates that Latrobe may have been planning to carry out this design in Madison's time.

List of Trees & Shrubs for the Presidents Garden.

		When to be procured
Large trees for Single trees & in lines —	Tulip tree	Saml. Davidson —
	Willow leaved oak	
	Elm.	
	Beech.	
	Ash.	S. Davidson
	Horse Chesnut.	Holt
	Walnut (English)	Hepburn
	Sugar maple	Billey
	Sycamore (English.	
	Larch.	
	Weeping willow	Holt
For Close plantation & Clumps. & for screens.	Mulberry – English, common Chenese	
	Catalpa	Wm. Taylor
	White, or Weymouth pine.	
	Hemlock	
	Mountain ash. (Rowentree).	S. Davidson
	Locust – common & Honey	Holt &c
	Spruce Pine.	
	Buckeye.	
	Redbud.	
	~~Chi~~ Chinese Arb. Vita	Mayne
	Golden willow	Holt
	Filbert.	Holt.
	Holley	

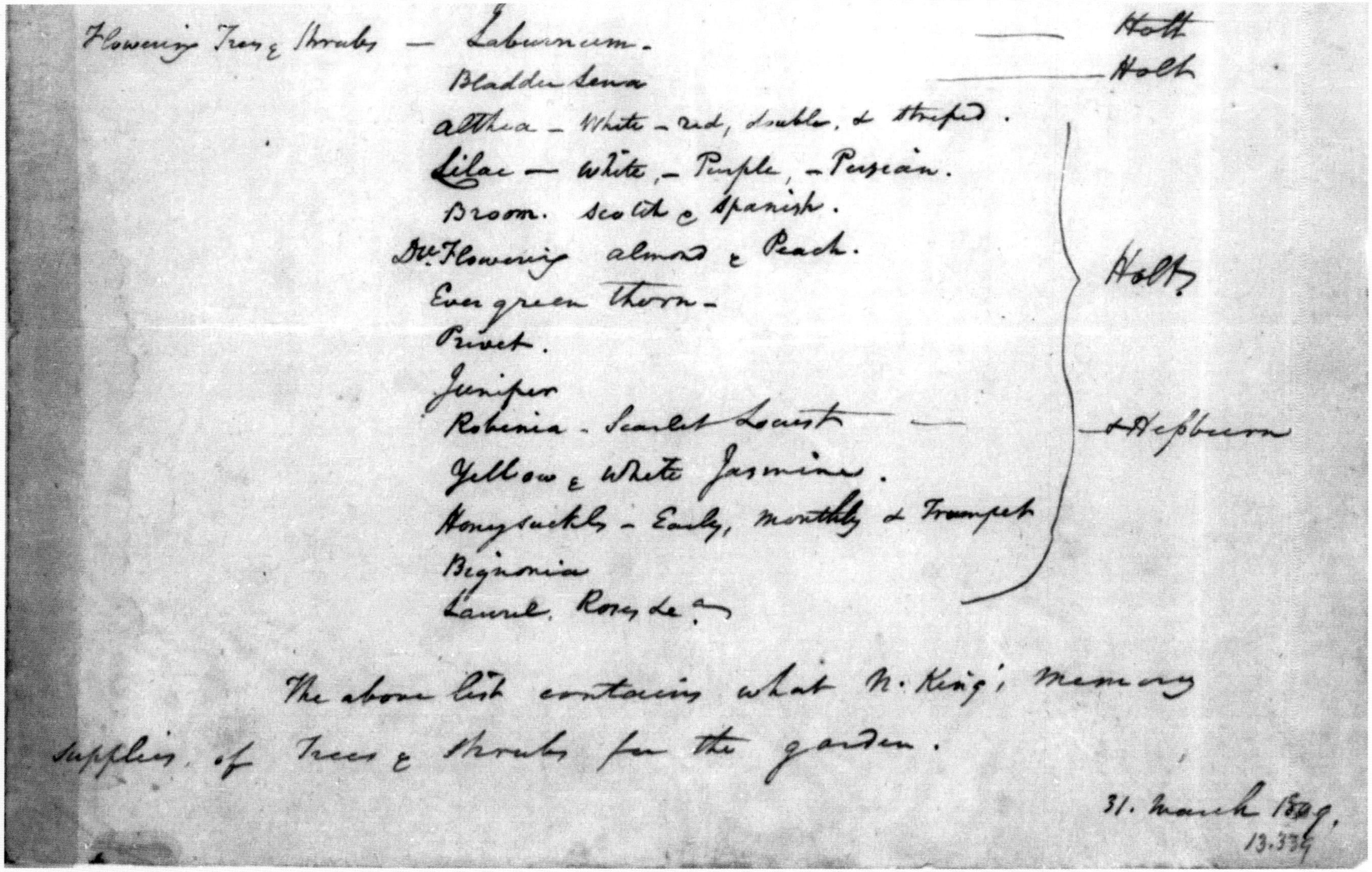

Flowering Trees & Shrubs — Laburnum. — Holt

Bladder Senna — Holt

Althea — White — red, double, & striped.

Lilac — White, — Purple, — Persian.

Broom. Scotch & Spanish.

Dw. Flowering almond & Peach.

Evergreen thorn —

Privet.

Juniper

Robinia - Scarlet Locust —

Yellow & White Jasmine.

Honeysuckles - Early, monthly & Trumpet

Bignonia

Laurel, Roses &c.

} Holt & Hepburn

The above list contains what N. King's memory supplies of Trees & shrubs for the garden.

31. March 1809.

13.339

Although the quantity of each type of tree or shrub is not given, the sheer number of species suggests that hundreds of plants would be needed. Two famous weeping willow trees were likely planted at this time, flanking Jefferson's monumental entrance arch at the southeast corner of the enclosure. These trees are the subject of a quip often attributed to Dolley Madison when, upon being congratulated on her new status as first lady, she jested, "I don't know that there is much cause for congratulation; the President of the United States generally comes in at the iron gate, and goes out at the weeping willows."

Thomas McGrath, White House gardener under Madison and the first to hold this position, probably took care of the actual planting, possibly under the supervision of Jean-Pierre Sioussat, Madison's chief steward. The work progressed rapidly, and on May 15, 1809, Joseph Dougherty, Jefferson's former coachman and assistant at the White House, wrote to the recently retired president: "Sir, if you were now at the President's house, you would scarcely know it. The north front is become a wilderness of shrubbery and trees." A similar scene likely unfolded on the South Grounds over the next five years. In 1813, 18-year-old William Preston, a future U.S. senator from South Carolina on his first visit to Washington, D.C., wrote in his journal that the White House and grounds were "very grand." This splendid view would not last long.

The United States declared war on Great Britain in 1812, and in August 1814 British troops marched on Washington and burned the White House. The fire gutted the mansion and left behind only the blackened exterior stone walls. Whether the grounds were similarly burned is not known for sure, but a painting from about 1820 very clearly shows thick plantings of tall ornamental trees rising above Jefferson's stone wall on the northeast side of the White House. Thus it seems likely that at least some of the trees and shrubs planted under Madison survived both the fire and subsequent rebuilding of the White House.

The reconstructed White House opened on January 1, 1818, but the repair work continued another two years, probably preventing any new planting on the grounds for up to six years following the fire. In 1819, President James

This painting by Anne Marguerite, Baroness Hyde de Neuville, c. 1820, shows thick foliage within the Jefferson stone wall on the North Lawn, indicating that at least some of the trees and shrubs planted during the Madison administration survived the White House fire of 1814.

John Quincy Adams would have been familiar with these illustrations from the horticulture books in his library: an oak tree illustrated in A Discourse of Forest Trees; thorn apple, small magnolia, common cranesbill, and mountain laurel from Jacob Bigelow's American Medical Botany; and a depiction of various germination stages from John Evelyn's Silva.

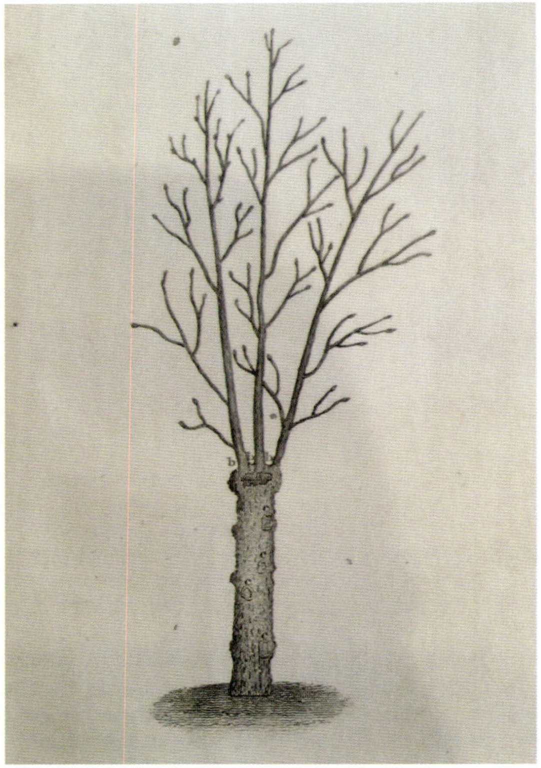

Monroe brought on a new White House gardener, Charles Bizet, who had previously worked for the Madisons at their Virginia home, Montpelier. Owing to a lack of surviving records, the full extent of Bizet's work at the White House is unknown, but he did plant additional shade trees and placed whitewashed wooden boxes around his young charges to protect them from the heels of clumsy workmen and the mouths of the president's cows and horses. Bizet most likely arranged the new plantings according to Charles Bulfinch's missing 1818 master plan for grading and ornamenting the landscape.

This tree-planting program continued during the presidency of John Quincy Adams, who took a great personal interest in the process. The younger Adams came to the White House an accomplished politician, having served as minister to the Netherlands and envoy to Portugal, Russia, Prussia, and Great Britain. He also became a Massachusetts state senator, U.S. senator, and secretary of state during the Monroe administration. While in this last position, Adams helped found Washington's first botanical garden, a forerunner to the current United States Botanic Garden on the National Mall. Since his youth he had nourished "a natural propensity" toward horticulture, taking particular pleasure in growing fruit and forest trees from seed. As he aged, Adams also developed a keen interest in botany, practical gardening, and forestry, but his political career left him little opportunity for these pursuits. He was 57 years old when he became president, and his knowledge of plants came almost entirely from books.

As president, Adams continued to consult these tomes. His voluminous diary, which he added to almost daily throughout his adult life, reveals that he often rose before dawn to study a chapter or two and frequently closed the day, long after nightfall, in much the same way. The books on his bedside table were not exactly light reading: John Loudon's *Encyclopaedia of Gardening* (1822) and *Encyclopaedia of Agriculture* (1825), each more than 1,200 pages, André Michaux's *North American Sylva* (1811–19), an expansive treatise on American forest trees, several of Henri-Louis Duhamel du Monceau's discourses on tree culture, and Jacob Bigelow's *American Medical Botany,* a

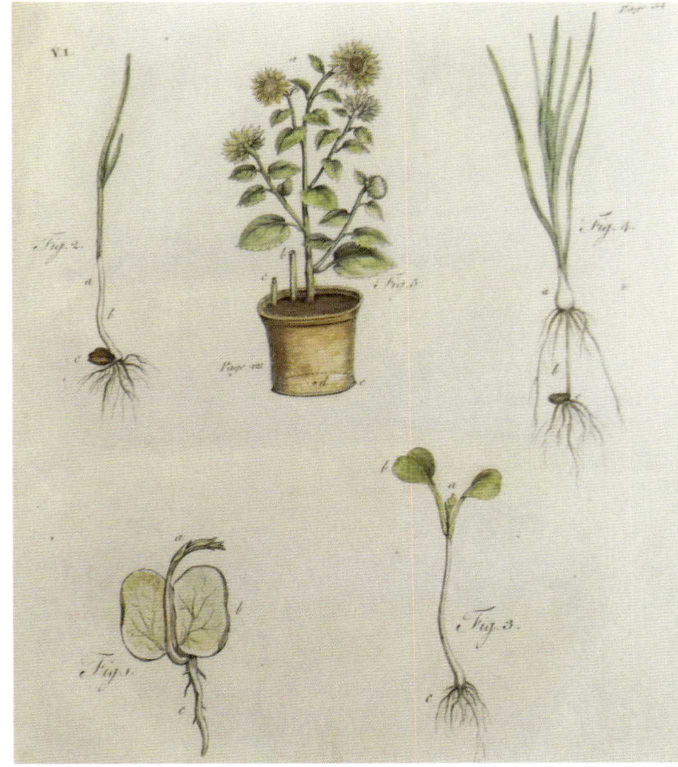

massive three-volume book published 1817–20 as the first comprehensive study of the medicinal properties of native plants. Adams even consulted John Evelyn's seminal 1664 masterwork *Sylva; or, A Discourse of Forest Trees*, one of the earliest published texts on forestry.

Adams's diary also records the exact date on which he chose to finally put his long-standing fascination with plants into practice and begin gardening in earnest: July 5, 1826. In an effort to identify three different species of oak growing on the White House grounds, he consulted his books but failed to find answers. Afterward, he wrote that the experience only served to remind him of "how much I have yet to learn." Simply reading about trees was no longer adequate, and with some trepidation Adams resolved to seek out his own answers on the President's Grounds. "I propose to commence a nursery the next autumn, and to plant acorns, hickory-nuts, and chestnuts. But what is the season for planting, and how is the nursery to be managed? I should have commenced this process at least thirty years since, but I have never had a permanent residence, and now I shall plant, if at all, more for the public than for myself."

Adams knew that the White House was not his permanent home and that he would likely be in the grave before his trees reached their full size. Still, he hoped that later generations would benefit from his efforts even if he would not. These trees were to be Adams's legacy at the White House, his gift to the American people, and he acted with a profound sense of responsibility toward the future. He privately admitted his hope that the trees he planted would outlast many more presidential administrations, but also admitted, "Who can look into futurity for the natural life of a row of oaks, and who can tell how soon they might be uprooted?"

President Adams established his tree nursery inside a plot of land enclosed by a post-and-rail fence a short distance southwest of the White House. Some trees grew from seed here, but many more were initially planted in beds southeast of the mansion and then transferred to the nursery as seedlings. Once his trees grew large enough, Adams had them moved to their permanent locations throughout the President's Grounds.

This painting from 1827 shows the tree nursery on the southwest side of the White House planted by President John Quincy Adams within a wooden post-and-rail fence.

Adams seems to have envisioned the entire White House landscape as one vast arboretum, to which he intended to add as many different species as possible. He successfully grew more than seven hundred saplings comprising more than twenty-five different types of trees: chestnut, Spanish chestnut, horse chestnut, English oak, chestnut oak, cork oak, post oak, white oak, willow oak, ash, ash-leaved maple, walnut, black walnut, shagbark hickory, shellbark hickory, maple, catalpa, pignut, tulip tree, white mulberry, almond, apricot, apple, cherry, lime, persimmon, and plum. As Adams described trees in general terms, referring only to "oaks" or "walnuts," this array may have been even more diverse. There was also a special type of peach that bore double-blossoms in the spring. The president credited his new gardener, John Ousley, with developing the variety, and as late as 1896 at least one of "Ousley's budded double-blossom peaches" still stood near the East Entrance.

Ousley came to the White House shortly after Adams's inauguration in 1825 and remained for nearly thirty years, serving as head gardener under nine presidents before finally departing in January 1852. His horticultural skills greatly impressed Adams, particularly his near-encyclopedic knowledge of plants. The gardener became the president's "nomenclator," patiently repeating the two-part Latin name for a staggering number of species again and again as Adams strove to commit them to memory. Ousley also provided much of the physical labor needed to plant and maintain all the president's new trees. When his schedule allowed, Adams helped, and he took great joy in digging the ground and personally planting seeds in the earth. The president also engaged in gardening as a scientific pursuit, analyzing soil samples, documenting plant growth over time, recording sprouting dates, and conducting experiments. He was particularly interested in studying whether seeds from grafted plants could successfully germinate.

Adams's experiments came to an abrupt end in 1829 following a failed bid at reelection against his greatest political foe, Andrew Jackson. But the new president also had a love of gardening, and the tree-planting program at the White House continued. Ousley still tended to the young trees in the nursery, but in the early 1830s, the government also began purchasing new

WASHINGTON SWEPT BY A HURRICANE

Great Damage Done by the Furious Gale Last Night.

WALLS OF HOUSES CRUSHED IN LIKE PAPER

Destruction of Property Reported in Every Part of the District.

FATALITIES REPORTED IN ALEXANDRIA

DAMAGED BY STORM

Many Trees Blown Down and Houses Unroofed.

MT. ZION CHURCH DEMOLISHED

Only One Serious Accident Was Reported.

EFFECT ON THE WIRES

Arriving as it did but a few hours after the announcement of the frightful tornado which caused destruction and havoc in St. Louis, the miniature cyclone which distributed itself throughout all Washington about 3:40 o'clock yesterday afternoon created probably as great an amount of fear in the hearts of the citizens as it did damage. The general opinion seemed to be that the St. Louis tornado had circled around and was making straight for the national capital. The storm was in evidence but a short time, but while it was here it displayed great activity and made itself deeply felt.

trees from commercial firms, including that of a local man named James Maher. "Jemmy" came to Washington from Ireland as a boy in 1798, his family forced to flee following an unsuccessful uprising against British rule. As a refugee, Maher grew up hating the British and came to idolize Jackson, "The "Hero of New Orleans," for his rousing victory at the close of the War of 1812. Jackson, in turn, found Maher loyal, lively, and amusing and chose to overlook his foul mouth and occasional drunkenness.

Jackson's favor probably explains why the federal government hired Maher to fill the post of "public gardener" for all of Washington, D.C. The job entailed developing the city's parks and public spaces and required the planting of large quantities of trees. In addition to earning a generous salary, Maher purchased his own stock whenever possible, thereby acting as both buyer and seller and earning money on each end of the deal. Many of the trees planted on the President's Grounds during Jackson's time came from Maher's nursery, but other orders arrived from the best suppliers on Long Island, then the center of the national gardening trade. Surviving receipts indicate that Maher procured upward of three hundred individual specimens comprising a variety of species still popular on the President's Grounds today: elms, oaks, ashes, European horse chestnuts, silver and sugar maples, red-twigged lindens, and American and European sycamores. Norway spruce trees were also planted on the grounds by 1840.

At the behest of President Millard Fillmore, the landscape gardener and "rural architect" Andrew Jackson Downing began to develop a plan for improving Washington's major public spaces in the early 1850s. Downing's death two years later prevented his design for the President's Grounds from moving beyond a rough sketch, but it is nonetheless clear that great numbers of trees figured prominently in his plans. Had Downing lived, an additional wave of tree planting most likely would have occurred. On the other hand, many preexisting trees would have had to be trimmed or removed entirely, particularly on the South Grounds, to provide a clear viewing lane down to the Washington Monument. Frederick Law Olmsted Jr. finally addressed this issue nearly a century later during the presidency of Franklin Roosevelt by opening up the South Vista, but Downing first envisioned thick groves of trees on either side of this principal sightline almost a century earlier.

As it turned out, two decades passed before tree planting resumed at the White House. When President Ulysses S. Grant expanded the South Grounds in the early 1870s, the new addition was in rough shape and largely devoid of vegetation. It became the head gardener's responsibility to populate this area with plants and trees. A self-made man named Alexander McKerichar currently held this position, and from 1871 to 1875 he supervised the planting of hundreds of shrubs, evergreens, and deciduous trees on the far South Grounds. "None but long-lived, hardy, and fine specimens" were selected, some sixty of which came from Lafayette Park and other public spaces around the city. McKerichar and his assistants took particularly good care of these transplants, securely tying them, keeping them well-watered, and applying a potent fertilizer made from a mixture of manure, lime, and "canal-muck." Owing to their efforts, only one of these relocated trees failed to take root on the President's Grounds.

President Rutherford Hayes moved into the White House in 1877 and soon hired an entirely new gardening staff. Henry Pfister, who first met Hayes in the president's hometown of Cincinnati, followed him to Washington and took over as head gardener. Pfister retained this position for the next thirty-five years, making him one of the longest serving gardeners in the history of the White House. Hayes kept Pfister and his assistants busy planting yet more trees, chiefly oaks, elms, spruces, firs, and horse chestnuts. Historic photographs show that many of these specimens were quite big, sometimes standing 10 feet tall or more when planted.

Several major storms during the 1890s damaged or destroyed some of the oldest and best trees outside the White House. The first struck in August 1893 and blew down three of the biggest trees on the South Grounds. One came up by the roots and brought with it an immense mound of earth. A second broke off just below ground level, leaving a gaping wound on the landscape. The third tree snapped like a twig about 20 feet up its trunk. After the storm, branches and other debris lay heaped about the grounds, and the *Baltimore Sun* reported that the loss of the trees seriously diminished the beauty of the White House. This damage, however, would seem tame compared to what came next.

In May 1896, a succession of three violent thunderstorms struck within days, one of which also brought a tornado with wind speeds in excess of 70 miles per hour. A Chicago newspaper pronounced it a "genuine Western cyclone." On the President's Grounds, it ripped up a pair of large trees and greatly damaged a "majestic and aged maple" described as the "finest tree in the park." Its branches were "twisted off and strewn in every direction." Many more trees were completely "denuded of limbs." The damage marked the path of the tornado, a "space of less than a hundred yards" across the President's Grounds. Still, the worst was yet to come.

In September, a disastrous hurricane hit Washington. The storm killed ten people, injured many others, and caused $50 million in property damage.

Within four months in 1896, two headlines in the Evening Star *reported severe storm damage to the White House Grounds. In May a tornado damaged many trees; in September a hurricane destroyed fifty trees and damaged 125 others.*

Two views of one of the three historic elms that were uprooted during a great storm of July 30, 1913, blocks the driveway on the North Lawn of the White House.

Outside the White House, the high winds destroyed fifty trees and seriously damaged at least 125 others, many of them historic specimens that had graced the grounds for decades. One was a maple on West Executive Avenue, perhaps the same maple damaged in the tornado four months earlier. Newspaper articles said it dated to Abraham Lincoln's presidency and that he planted it himself. Relic hunters descended on the scene and carried away broken branches and bits of bark. The White House also received many requests for wood from the tree to make canes, and Pfister and his gardeners did their best to comply. Today collectors treasure canes made from this "Lincoln Tree."

On the North Grounds trees had "smashed" through the iron perimeter fence and underlying stone coping. Flying debris also collided with one of the stone gateposts. After the storm it took seventeen days to clear the grounds of limbs and litter. Workmen carted away nearly five hundred wagon loads of wood and brush, and the task of cutting back the badly disfigured trees stretched into winter. Even so, about ten mangled trees remained as the storm had left them; there was not enough money or manpower for all the repairs. The gardeners planted seventy new trees and shrubs in thirty-five varieties to replace those that were lost.

Despite the storms, the landscape of the President's Grounds remained densely packed and increasingly cast in shade as the trees grew ever larger. The gardeners do not appear to have trimmed the trees very often, and period photographs of the South Grounds show trees almost everywhere, planted apparently at random. In an age when the public could still come and go at will, these thick groves provided a measure of privacy for the residents of the White House. They also gave the grounds a picturesque aura, with dappled light filtering down through branches and leaves. The tree plantings on the North Grounds were somewhat less prolific and more formal. Many of these trees were arranged more or less in straight-line rows. Evidence of this design is still discernible today and reflects the role of the North Grounds as the public front lawn of the White House.

The trees continued to grow but otherwise the scene changed little until 1935 when Frederick Law Olmsted Jr. submitted his plan for improving the

A row of new trees planted on the North Lawn, c. 2014, are positioned to preserve the primary sightline of the North Portico.

landscape. He held the grand, old trees in extremely high regard and ranked them, along with the topography and the mansion itself, among the site's most outstanding characteristics. While he recognized the "need for planting a very few long-lived trees carefully chosen as to species for each location," he also felt that, generally, the grounds would benefit from having fewer trees, not more. There were already many underdeveloped, partly obscured specimens crowded beneath the taller, larger trees. The goal had apparently been to have the young trees grow up beneath the canopy and then succeed the older ones when they died off. In many cases this arrangement was species specific—an elm planted under an elm, a poplar under a poplar—so that the overall appearance would remain the same even as the trees themselves changed. Such a plan seemed logical in theory, but it failed in practice. Olmsted dismissed it as an "utterly futile" idea, completely inappropriate for such a highly regarded ornamental landscape as the President's Grounds. Given the space constraints and lack of sunlight, the young trees would certainly be "suppressed and worthless before they are needed." Meanwhile, overcrowding diminished the pleasant appearance of the larger, more established trees, and in Olmsted's mind these trees mattered most.

If previous tree-planting programs emphasized quantity, Olmsted instead focused on quality. He called for a thinning out of the understory and a moratorium on planting new trees except when filling in undesirable gaps within groves or replacing pre-existing specimens that had died or been otherwise removed. He deemed saplings and other young, immature trees unsuitable, detracting from the impression made by the older trees, and he went so far as to propose a minimum age requirement of fifty years for all new plantings. While this latter recommendation has not been followed to the letter, Olmsted's other views on tree planting and management have generally been adhered to.

Dealing with the trees on the North Grounds proved relatively simple and straightforward. The North Front of the White House had always been its public face. Almost from the very beginning, it was meant to be seen, and therefore the trees and other plantings on the North Grounds had never really obscured the mansion, especially not along the primary sightline to-

ward the North Portico. In a few places it would be necessary to pare down or simplify the plantings so as to create, as Olmsted put it, "a more dignified and less restless and self-assertive appearance." But these changes would be minor, amounting to only a slight modification of the existing landscape. As a result, all Olmsted really had to do was stress that the current axial relationship between the White House and Lafayette Park should be maintained. Things were not so simple to the south.

The South Grounds had been more densely planted throughout the nineteenth century and consequently required a more thorough thinning out. Olmsted further argued that the mass plantings had unduly impinged on the all-important South Vista toward the Washington Monument, "choking up and frittering away" this iconic view by blocking it with trunks, branches, and leaves. Opening up this sightline meant that even more trees would have to be eliminated or transplanted. Making the case to remove that many trees was difficult, but after some initial opposition his recommendations were adopted, and the appearance of the presidential landscape improved. The South Grounds were cleared, except for the stately groves of trees along the sides, which now served to screen views from the surrounding streets. Opening up the understory also gave the place a brighter, more cheerful appearance and allowed the old, majestic specimens to be better appreciated.

Today the President's Grounds resemble a natural woodland. This happy result is due largely to Olmsted's vision, which still forms the basis for all work carried out on the landscape. Before a new tree is added to the site, great care is taken to make sure that its type and location harmonize with its surroundings. The trees define the White House Grounds. More than just decorative features, they give the space structure and meaning. Because of their longevity, the trees also impart continuity and permanence over time. They stand as silent sentinels and living witnesses to American history.

On the south side of the White House, groves of mature trees provide privacy while the open vista toward the Washington Monument is preserved as Frederick Law Olmsted intended.

The Oldest Trees

IN 1900, HEAD GARDENER Henry Pfister took a long walk around the President's Grounds. With pencil and paper, he marked the precise location of every tree, large shrub, and climbing vine within the fence, and he identified each by species. In all there were 472 individual specimens, about the same as the 500 or so located on the grounds in modern times. Overlaying Pfister's map with the current White House planting plan reveals that fewer than fifty trees have survived from his time until now, meaning that a little less than 10 percent of the present tree population dates from the nineteenth century.

Detail of a grafted Camperdown Elm on the White House Grounds.

A chain helps stabilize a large old linden tree on the North Lawn.

Pfister's is the earliest inventory, so it is hard to know how old these trees really are. Nursery receipts and plant lists for the White House in the National Archives and Library of Congress are not enough to confirm that any of the trees present on the grounds today are the ones listed. Only an analysis of growth rings, after these trees die, would be able to ascertain their age. But in the meantime, it is possible to make some educated guesses.

In 1991, a venerable American elm planted atop the eastern mound on the South Grounds died. Its health had been visibly declining for decades; it was struck by lightning in 1965 and damaged in a 1976 storm. Superintendent of Grounds Irvin Williams determined that it was 165 years old and that it dated to 1826. The elm had probably been planted by John Ousley or perhaps even President John Quincy Adams himself. The tree looked every bit its age, and there are a handful of other trees remaining on the White House Grounds that appear similarly ancient.

Of the possible candidates, one in particular stands out: a massive chestnut oak southeast of the mansion. Even to the casual observer, it is obviously

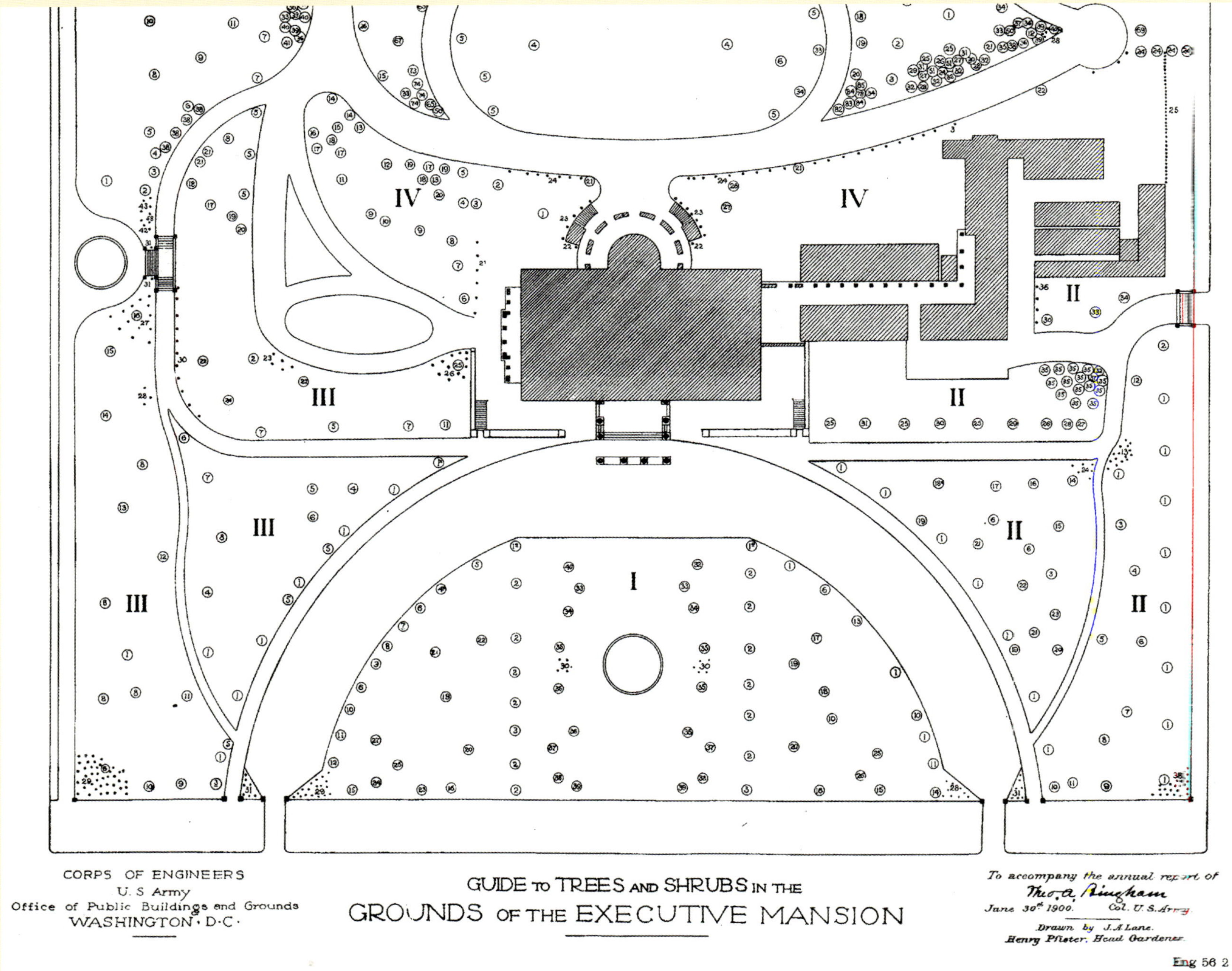
IV
IV
III
III
III
II
II
II
II
I
CORPS OF ENGINEERS
U. S Army
Office of Public Buildings and Grounds
WASHINGTON · D·C·
GUIDE TO TREES AND SHRUBS IN THE
GROUNDS OF THE EXECUTIVE MANSION
To accompany the annual report of
Theo. A. Bingham
June 30th 1900. Col. U. S. Army
Drawn by J. A. Lane.
Henry Pfister, Head Gardener.
Eng 56 2

List of trees and shrubs in the grounds of the Executive Mansion.

[Compiled by Mr. Henry Pfister, Head Gardener.]

[DESCRIPTION: T. for tree; S. for shrub; E. for evergreen tree; E. S. for evergreen shrub; and V. for vines. The numbers refer to locations on map.]

SECTION I.

No.	Description.	Name.	No.	Description.	Name.
1	T.	American elm.	20	T.	Acacia nemu. (Japan acacia).
1°	T.	American elm, planted March, 1878, by President R. B. Hayes.	21	T.	Quercus concordia (Golden-leaved oak).
2	T.	European linden.	22	T.	Sugar maple.
3	T.	American linden.	23	T.	American ash.
4[x]	T.	Sweet gum, planted Apr. 6, 1892, by President Benjamin Harrison.	24	S.	Magnolia halleana,
5	T.	Koelreuteria japonica.	25	S.	Peach (doubled red flowered).
6	T.	Salisburia adiantifolia, maiden-hair tree.	26	S.	Magnolia species.
7	T.	Scarlet oak, planted Apr. 6, 1892, by Benjamin H. McKee, grandson of President Harrison.	27	E. S.	Boxbush (or boxwood).
8	T.	Magnolia macrophylla.	28	E. S.	Bambus (clump of 10).
9	T.	Acer. tartaricus ginnala.	29	S.	Desmodiums, lilacs, and Japan maples in a clump.
10	T.	Horse chestnut.	30	S.	Japanese quinces, forsythias, and Hydrangeas paniculata in a clump.
11	S.	Halesia tetraptera (silver bell).	31	S.	Lagerstræmias. forsythias, double mock oranges and althæas.
12	T.	Red oak.	32	E.	Picea cephalonica.
13	T.	European beech.	33	E.	Picea nordmanniana.
14	T.	Silver-leaved poplar.	34	E.	Pinus excelsa.
15	T.	Scarlet maple.	35	E.	Abies orientalis.
16	T.	English ash.	36	S.	Prunus pissardii.
17	T.	Buckeye (yellow).	37	E.	Abies orientalis compacta.
18	T.	Swamp maple.	38	T.	Magnolia soulangeana.
19	T.	Tulip poplar.	39	S.	Weigelia rosea.
			40	E.	Abies amabilis.

SECTION II.

No.	Description.	Name.	No.	Description.	Name.
1	T.	American elm.	20	T.	Horse-chestnut.
2	T.	Acacia nemu (Japanese acacia).	21	T.	American ash.
3	T.	Mountain ash.	22	T.	English ash.
4	T.	Tulip poplar.	23	T.	Buckeye (yellow flowered).
5	T.	Swamp maple.	24	E. S.	Yuccas in a clump.
6	T.	Sugar maple.	25	T.	Cut-leaved beech.
7	T.	European oak.	26	T.	Fern-leaved beech.
8	T.	American black oak.	27	T.	Cut-leaved weeping birch.
9	T.	Rivers' (purple leaved) beach.	28	S.	Magnolia halleana.
10	T.	Scarlet maple.	29	E.	Abies (Picea) polita.
11	S.	Cytisus laburnum (golden chain).	30	E.	Abies amabilis.
12	E. S.	Bambus (golden).	31	E.	Abies douglasii.
13	E. S.	Azalea amæna and hypericum moserianum, in a clump.	32	S.	Paeonia (tree).
14	T.	American linden.	33	E.	Abies nobilis.
15	T.	European linden.	34	E.	Abies canadensis compacta glauca.
16	T.	Buckeye (red flowered).	35	S.	Japanese maples.
17	T.	Silver-leaved poplar.	36	V.	Lonicera halleana (Hall's honeysuckle.
18[x]	T.	Scarlet oak, planted March 18, 1898, by President McKinley.	37	E. S.	Boxwood (tree).
19	T.	Soft or silver maple.	38	S.	Deutzia giaulis pæonias. lilacs, and yuccas in a clump.

SECTION III.

No.	Description.	Name.	No.	Description.	Name.
1[x]	T.	American elm, planted March, 1880, by Scott Hayes, youngest son of President Hayes.	17	E.	Pinus excelsa.
1	T.	American elm.	18	E.	Thuia occidentalis aurea.
2	T.	Fern-leaved beech.	19	E.	Taxus baccata washingtoni.
3	T.	Scarlet oak, planted April 6, 1892, by Marie L. McKee, grand daughter of President Benjamin Harrison.	20	E. S.	Aucuba variegata.
4	T.	American ash.	21	S.	Prinos verticilatus (black alder).
5	T.	Soft or silver maple.	22	E. T.	Magnolia grandiflora.
6	T.	Sugar maple.	23	E.	Thuia siberica in a clump.
7	T.	American sycamore.	24	T.	Horse-chestnut (double flowered).
8	T.	English ash.	25	S.	Halesia hispida.
9	S.	Cytisus laburnum (golden chain).	26	S.	Smoke tree, hypericums, lilac, cornus, viburnum, and lonicera in a clump.
10	T.	Rivers's (purple-leaved) beech.	27	S.	Japanese maples, althæas, and double-flowering almonds in a clump.
11	T.	European linden.	28	E. S.	Limonia trifoliata (hardy orange).
12	T.	Horse-chestnut.	29	S.	Tree pæonia, herbaceous pæonias, althæas, deutzia gracilis, lagerstræmia, and lilacs in a clump.
13	T.	Tulip poplar.	30	V.	Clematis paniculata.
14	T.	Acer platanoides (sycamore maple).	31	V.	Bignonia grandiflora, large-flowered trumpet vine.
15	T.	Paulownia imperialis.			
16	S.	Dimorphanthus mandschuricus.			

SECTION IV.

No.	Description.	Name.	No.	Description.	Name.
1	E. S.	Boxwood.	16	S.	Japanese quince.
2	T.	Acer platanoides (sycamore maple).	17	S.	Spirea prunifolia.
3	V.	Wistaria japonica (white flowered).	18	S.	Spirea reevesii.
4	S.	Magnolia glauca.	19	S.	Forsythia fortunii.
5	E.	Taxus baccata pyramidalis.	20	S.	Spirea reevesii flora plena.
6	S.	Syringa Ligustrina.	21	V.	Rose (Empress of China).
7	S.	Peach (double white flowered).	22	V.	Wistaria sinensis.
8	S.	Cornus mas (Cornelian cherry).	23	V.	Honeysuckle (Hall's and red flowered).
9	S.	Peach (double red flowered).	24	V.	Clematis and roses.
10	E.	Pinus strobus compactus nanus.	25	E.	Thuia siberica (Hedge).
11	T.	Purple-leaved beech.	26	E. T.	Magnolia grandiflora (Callisonier var.).
12	T.	Horse-chestnut.	27	E. T.	Magnolia grandiflora (ordinary var.).
13	S.	Mock orange.			
14	S.	Lilac (white flowered).			
15	S.	Lilac (Persian).			

This "Guide to Trees and Shrubs in the Grounds of the Executive Mansion" by J. A. Lane and Henry Pfister, c. 1900, identifies the location of each tree with a dot that is keyed to a planting list specifying its species. The sections marked I through IV include areas closest to the house on the north, east, south, and west.

This portion of the "Guide to Trees and Shrubs in the Grounds of the Executive Mansion" by Lane and Pfister includes the planting list for the trees in large areas of the South Grounds, sections marked V through VII.

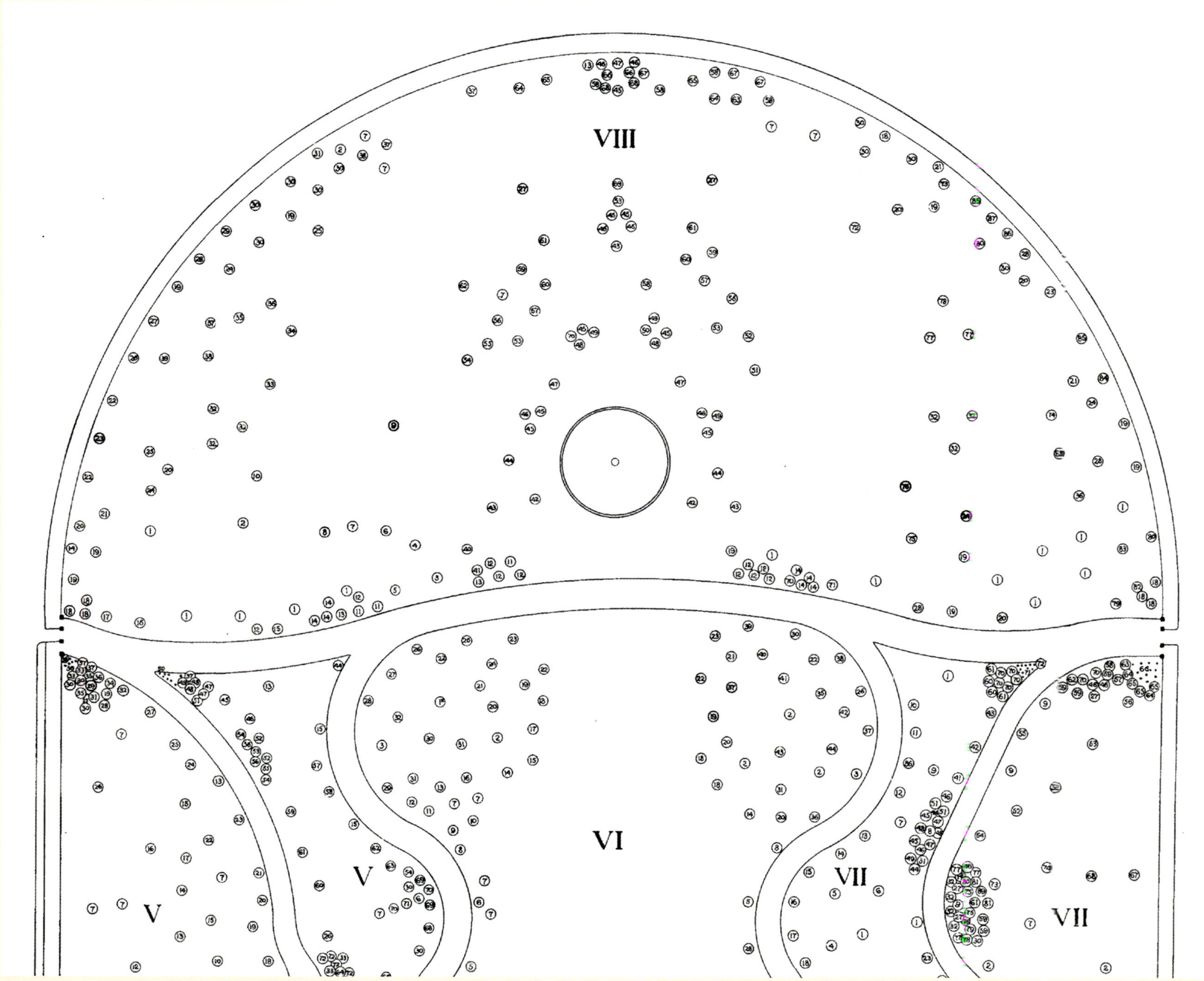

List of trees and shrubs in the grounds of the Executive Mansion—Continued.

[DESCRIPTION: T. for tree; S. for shrub; E. for evergreen tree; E. S. for evergreen shrub; and V. for vines.]

SECTION V.

No.	Description.	Name.	No.	Description.	Name.
1	T.	Paulownia imperialis.	39	E. S.	Box. Green foliage.
2	E.	Libocedrus decurens (red wood).	40	S.	Berberis purpurea (purple leaved).
3	E.	Abies douglasii.	41	S.	Spirea reevesii flora plena.
4	E.	Abies pungens (Colorado blue spruce).	42	S.	Hypericum moserianum.
5	E. T.	Magnolia grandiflora.	43	S.	Desmodium japonicum (white and purple variety).
6	E.	Picea nordmanniana.	44	T.	American sycamore.
7	T.	Soft or silver maple.	45	E.	Abies (Picea) polita.
8	T.	Spanish chestnut.	46	E.	Abies pinsapo.
9	T.	Cedrella sinensis.	47	E.	Cedrus deodara glauca.
10	T.	Cork-barked oak.	48	E.	Thuiopsis standishii.
11	T.	Pin oak.	49	E.	Cupressus lawsoniana.
12	T.	White oak.	50	E.	Yucca.
13	T.	Black oak.	51	S.	Magnolia Lenne.
14	T.	Japanese oak (Daimio).	52	T.	Crab apples (Siberian).
15	T.	Horse-chestnut.	53	S.	Ptelea trifoliata aurea.
16	T.	Black walnut.	54	S.	Lagerstraemia, rosea, alba, and purpurea.
17	T.	Acer negundo (ash-leaved maple).	55	S.	Weigelia rosea.
18	T.	Acer platanoides (sycamore maple).	56	S.	Spirea reevesii.
19	S.	Japanese red bud (Judas tree).	57	S.	Cornus mas. (Cornelian cherry).
20	S.	Cerasus jap. rosea pendula (weeping cherry).	58	T.	Carpinus betulus (European hornbeam).
21	T.	Cut-leaved beech.	59	T.	American ash.
22	T.	American beech.	60	T.	English ash.
23	S.	American red-bud (Judas tree).	61	S.	Diospyrus lotus (European persimmon).
24	T.	Acer schwedlerii (red-leaved maple.)	62	E.	Pinus excelsa.
25	S.	Cerasus acida (sour cherry).	63	T.	Tulip poplar.
26	T.	Sugar maple.	64	E.	Picea cephalonica.
27	T.	Koelreuteria japonica.	65	T.	Buckeye (yellow flowered).
28	S.	Chionanthus virginicus fringe tree).	66	T.	Larix europaea (European larch).
29	S.	Exochorda grandiflora.	67	S.	Snowball.
30	S.	Lilac (Persian).	68	S.	Magnolia halleana.
31	S.	Calycanthus floridus (sweet shrub).	69	S.	Spirea prunifolia.
32	S.	Althæa rosea.	70	S.	Jasminium nudiflorum.
33	S.	Mock orange.	71	S.	Japanese cherry.
34	S.	Mock orange (double flowered).	72	S.	Deutzia crenata.
35	S.	Ribes aurea (Missouri currant).	73	S.	Lilac (white flowered).
36	S.	Almond (double pink flowered).	74	S.	Lilac (common purple).
37	E. S.	Box. Yellow foliage.			
38	S.	Paeonia (tree),			

SECTION VI.

No.	Description.	Name.	No.	Description.	Name.
1x	T.	American elm, planted by John Quincy Adams.	22	S.	Cornus florida flora rubra (red-flowered dogwood).
2	T.	Horse-chestnut.	23	T.	Acacia nemu. (Japanese acacia.)
3	T.	Rivers's (purple leaved) beech.	24	S.	American red bud (Judas tree).
4	E. S.	Box tree (Buxus sempervirens).	25	T.	Tulip tree, with variegated foliage.
5	T.	Soft or silver maple.	26	S.	Hammamelis japonica (Japanese witch hazel).
6	T.	Sugar maple.	27	T.	Willow (Russian).
7	S.	Magnolia purpurata.	28	T.	Norway maple.
8	T.	Platanus orientalis (Oriental sycamore).	29	T.	Cherry (double yellow flowered).
9	S.	Peach (double white flowered).	30	T.	Koelreuteria paniculata.
10	T.	Horse-chestnut (double flowered).	31	T.	European beech.
11	S.	Styrax japonica,	32	T.	Crab apple (Siberian).
12	E.	Picea nobilis.	33	T.	Tulip poplar.
13	E.	Abies pungens (Colorado blue spruce).	34	T.	Purple beech (weeping).
14	E.	Pinus cembra compacta (Swiss mountain pine).	35	T.	Virgilia lueta (yellow wood).
15	E.	Pinus coraeensis (Corean pine).	36	S.	Cerasus sieboldii, flora rosea plena (double pink-flowered cherry).
16	E.	Cerasus avium alba plena (double white cherry).	37	T.	Acer wierii (cut-leaved maple).
17	E.	Picea pectinata species.	38	T.	Cut-leaved beech.
18	E.	Pinus excelsa (Himalayan pine).	39	T.	Cut-leaved birch (European).
19	E.	Cedrus atlantica glauca.	40	T.	Paulownia imperialis.
20	E.	Pinus austriaca (Austrian pine).	41	T.	English ash.
21	E.	Pinus strobus (white pine).	42	S.	Peach (double red flowered).
			43	T.	Catalpa bignonoides.
			44	T.	Acer negundo (ash-leaved maple).

SECTION VII.

No.	Description.	Name.	No.	Description.	Name.
1	T.	Soft or silver maple.	43	T.	Koelreuteria paniculata.
2	T.	Horse-chestnut.	44	S.	Exochorda grandiflora.
3	T.	Quercus concordia (golden-leaved oak).	45	S.	Almond (double rose flowered).
4	T.	Gleditschia sinensis (honey locust).	46	S.	Mock orange.
5	T.	Scarlet oak.	47	S.	Lagerstraemia rosea.
6	T.	Laurel-leaved oak.	48	S.	Jasminium nudiflorum.
7	T.	Rivers's purple beech.	49	S.	Spirea fortunii.
8	T.	American locust.	50	S.	Hydrangea Thomas Hogg.
9	T.	Weeping silver linden.	51	S.	Spirea reevesii flora plena.
10	T.	Acer ritenbachi (dark-leaved maple).	52	T.	Black oak.
11	T.	Carpinus betulus (European hornbeam).	53	T.	Ulmus camp. pendula camperdownii (Camperdown weeping elm).
12	E.	Pinus excelsa.	54	T.	American linden.
13	S.	Cornus florida (dogwood).	55	T.	Cut-leaved beech.
14	T.	Scarlet maple.	56	T.	Sugar maple.
15	T.	Phellodendron amurense (Chinese cork tree).	57	S.	Chionanthus virginica (fringe tree).
16	T.	Catalpa bignonoides	58	S.	Hydrangea paniculata.
17	S.	Syringa japonica (Jap. tree lilac).	59	S.	Deutzia crenata flora plena.
18	T.	Sweet gum (liquidambar styracfl).	60	S.	Amorpha fruticosa (false indigo).
19	T.	Crab apple (Siberian).	61	S.	Spirea prunifolia.
20	T.	English walnut.	62	S.	Forsythia fortunii.
21	T.	Spanish chestnut.	63	S.	Acer poly. jap. disectum purpureum.
22	T.	Tulip poplar.	64	S.	Acer poly. jap. atrapurpureum.
23	E. S.	Bambusa metake.	65	S.	Paeonia (tree).
24	E. S.	Boxwood.	66	E. S.	Yucca in a clump.
25	S.	Ligustrum ovalifolium (California privet).	67	T.	English elm.
26	S.	Halesia hispida.	68	T.	Silver-leaved poplar.
27	S.	Japanese quince.	69	T.	Paulownia imperialis.
28	E. S.	Mahonia japonica.	70	S.	Lilac (purple, large flowered).
29	S.	Hypericum prolificum.	71	S.	Duetzia gracilis.
30	S.	Mock orange (double flowered).	72	E. S.	Yucca filamentosa.
31	S.	Spirea thunbergii.	73	T.	Black walnut.
32	S.	Paeonia herbaceous.	74	T.	American sycamore,
33	E.	Cedrus atlantica.	75	S.	Lilac (Persian).
34	E.	Abies orientalis.	76	S.	Althaea variegata.
35	E.	Abies (Picea) polita.	77	S.	Ceanotus (gloir de plantier).
36	S.	Weigelia rosea folis variegatis.	78	S.	Lagerstraemia alba.
37	E.	Picea nordmanniana.	79	S.	Calycanthus floridus (sweet shrub).
38	E.	Abies pungens (Colorado blue spruce).	80	S.	Weigelia rosea.
39	S.	Lagerstraemia violacea.	81	S.	Althaea rosea.
40	S.	Rosa rugosa.	82	S.	Limonia trifoliata (hardy orange).
41	T.	Betula alba (European birch).	83	E. S.	Boxwood (tree).
42	T.	Acacia nemu (Jap. acacia).	84	S.	Spirea, reevesii.
			85	E.	Thuia occidentalis pyramidalis.
			86	T.	Acer ginnala.

SECTION VIII.

No.	Description.	Name.	No.	Description.	Name.
1	T.	American elm.	23	T.	American ash.
2	T.	Salisburia adiantifolio (maidenhair tree).	24	T.	Field maple.
3	T.	Willow-leaved oak.	25	T.	Sophora japonica (Japanese varnish tree).
4	T.	Laurel-leaved oak.	26	T.	Scarlet maple.
5	T.	Acer schwedlerii.	27	T.	Paulownia imperialis.
6	E.	Thuia siberica.	28	T.	Acer negundo (ash-leaved maple).
7	E.	Abies excelsa.	29	T.	English ash.
8	E.	Picea nordmanniana.	30	T.	Horse chestnut.
9	T.	Ulmus campestris pendula camperdownii (Camperdown weeping elm).	31	T.	Virgilia lutea (yellow wood).
10	E.	Thuia occidentalis.	32	T.	Swamp cedar.
11	S.	Mock orange.	33	E.	Thuia occidentalis filifera.
12	S.	Lilac (common purple).	34	T.	Olive oak.
13	E. S.	Boxwood.	35	T.	Cork-barked oak.
14	S.	Althæa rosea.	36	T.	Black oak.
15	T.	Wild cherry.	37	E.	Abies canadensis (hemlock).
16	T.	Plantaus orientalis (Oriental sycamore).	38	E.	Pinus scoticus (Scotch pine).
17	T.	Silver-leaved linden.	39	S.	Andromeda arborea (sorrel tree).
18	S.	Rhus cotinus (smoke tree).	40	T.	Scarlet oak.
19	T.	Sugar maple.	41	E.	Virginian cypress.
20	T.	Koelreuteria paniculata.	42	T.	Fagus sylvaticus pendula (weeping beech.
21	S.	Cytisus laburnum (golden chain).	43	S.	Tamarix africana.
22		Norway maple.	44	T.	Pyrus malus parkmanni (large-flowered apple tree).
			45	S.	Weigelia rosea.

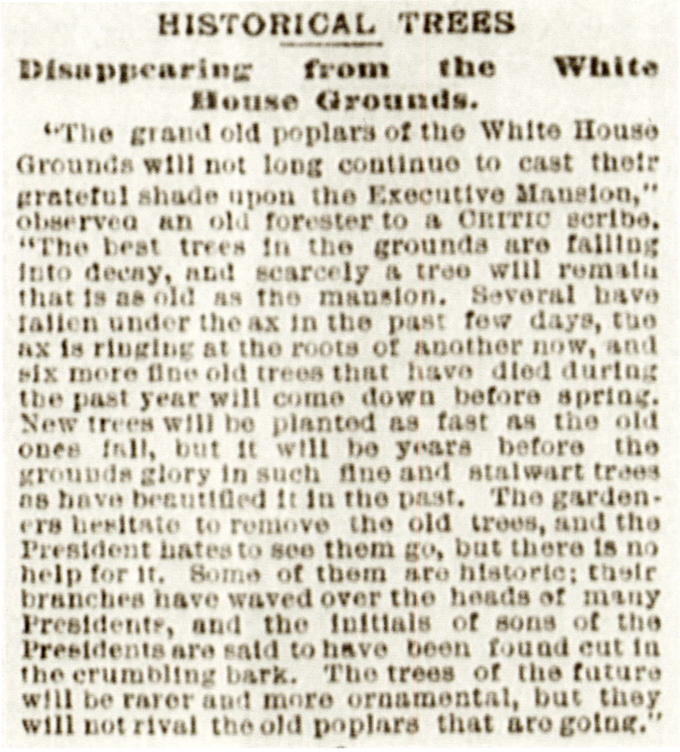

HISTORICAL TREES

Disappearing from the White House Grounds.

"The grand old poplars of the White House Grounds will not long continue to cast their grateful shade upon the Executive Mansion," observed an old forester to a CRITIC scribe. "The best trees in the grounds are falling into decay, and scarcely a tree will remain that is as old as the mansion. Several have fallen under the ax in the past few days, the ax is ringing at the roots of another now, and six more fine old trees that have died during the past year will come down before spring. New trees will be planted as fast as the old ones fall, but it will be years before the grounds glory in such fine and stalwart trees as have beautified it in the past. The gardeners hesitate to remove the old trees, and the President hates to see them go, but there is no help for it. Some of them are historic; their branches have waved over the heads of many Presidents, and the initials of sons of the Presidents are said to have been found cut in the crumbling bark. The trees of the future will be rarer and more ornamental, but they will not rival the old poplars that are going."

This clipping from the Washington, D.C., newspaper, Evening Critic, *highlights the sad loss of several tulip poplars from the White House Grounds, January 7, 1884.*

very tall and large and therefore probably very old. Chestnut oak is also a long-lived, slow-growing species. Its trunk measures more than 5 feet in diameter, making it a true giant. Other chestnut oaks with similar trunk diameters have been proven to be more than three hundred years old, and the maximum life span for the species exceeds four-hundred years. Although judging a tree's age from its size is far from an exact science, this line of reasoning indicates that the chestnut oak on the President's Grounds is very old indeed, perhaps even older than the mansion itself.

That the chestnut oak on the President's Grounds is this old is at least plausible. The government's contract with the previous landowners allowed them to "freely cut down, take, and carry away, and use" trees on their former property during construction of the White House, except those "trees and wood growing as the President or commissioners aforesaid may judge proper, and give notice, shall be left for ornaments, for which the just and reasonable value shall be paid." Subsequent records make it clear that the commissioners invoked this privilege on at least two occasions. Shortly thereafter, Benjamin Henry Latrobe noted a "Great Walnut" on the South Grounds that he used as a fixed reference point while carrying out Thomas Jefferson's landscape improvements. Nearly seventy years later, a local newspaper article bemoaned the loss of several tulip poplars described at the time as being "the best trees on the grounds" and "as old as the mansion." The article suggests that the poplars were not the only trees to have been established naturally before the building of the White House, but that most had already died off.

Perhaps the chestnut oak is the last such survivor. If so, it is doubtless the oldest tree on the President's Grounds and probably among the oldest in all of Washington, D.C. It would also hold the unique distinction of being the only living thing to have seen the comings and goings of every single president of the United States. Taking a core sample from its trunk could let us know for sure, but potentially do major harm to the tree as well. So we will simply have to be patient and wait for it to live out its natural life; only in death will it give up its secret.

A detail of cork tree on the South Lawn, reveals its unusual bark, which may be an indication of advanced age.

A group of bald cypress on the South Lawn in winter.

A view from above of great elms along the North Drive.

A group of bald cypress on the South Lawn.

A red Japanese maple is seen to the left of a large group of hollies on the South Lawn.

A tree lilac specimen is seen here, just south of the putting green.

Horse chestnut

Camperdown elm

Commemorative Trees

COMMEMORATIVE TREE PLANTING has a long and storied history at the White House. There are currently more than three dozen trees on the grounds planted by the presidents and first ladies to mark their time in office. The present collection represents seventeen different presidential administrations, and many of the spades used in the ceremonial shoveling of dirt have also been preserved for posterity.

The spades used by presidents and first ladies to plant commemorative trees on the White House grounds bear placques recording the details of each planting.

The Japanese maple in the foreground was planted by Frances Folsom Cleveland in 1893. The smaller specimen on the far right was planted by Rosalynn Carter in 1978 to replace another maple planted by Mrs. Cleveland that had died.

The oldest surviving tree with a proven association with a resident of the White House is a diminutive threadleaf Japanese maple planted by Frances Folsom Cleveland on the far South Grounds in 1893. Until recently, however, this honor went to a scarlet oak planted near the Northeast Gate by Benjamin Harrison and his grandchildren in 1889; it was removed in 2009 after 120 years on the grounds. Most recently, President Barack Obama and First Lady Michelle Obama have continued the long-standing tradition, planting a little-leaf linden tree on the North Grounds on October 28, 2009.

John Quincy Adams inaugurated the custom of planting commemorative trees but it ebbed shortly after his presidency came to a close in 1829. Rutherford B. Hayes reintroduced the practice in the late 1870s, planting many commemoratives, including Ohio buckeyes (horse chestnuts) in honor of his native state. He further intended to plant a tree for each previous president as well as a tree for each state in the Union. Although he did not realize this larger goal, he generated enough interest in the idea that almost all his successors planted at least one tree on the grounds during their time at the White House.

Unfortunately, many of the commemorative trees have not survived. This absence does not make their histories any less compelling, however. A few of the most interesting of these lost trees are described here.

John Quincy Adams's Historic Trees

When he began gathering stock for his tree nursery at the White House, John Quincy Adams sought out descendants of trees associated with important people and events in American history. From Belmont, the Peters family estate outside of Philadelphia, Richard Peters Jr. agreed to supply him with both nuts and young trees descended from a Spanish chestnut that had been planted by George Washington just before the close of his presidency in honor of his friendship with Peters's father, a delegate to the Continental Congress and a captain in the Continental Army during the American Revolution. On April 24, 1827, Adams planted three of these "Peters Chestnuts" outside the White House. The following October, Adams personally collected six white oak acorns from a Baltimore tree riddled with more than twenty rounds of ammunition fired by British soldiers during their assault on the city during the War of 1812. He added them to his tree nursery, and on April 29, 1828, noted that one had sprouted.

Famous Trees of Washington

The John Quincy Adams Elm.

On a rounded knoll southeast of the White House stands a graceful elm, towering over all the other trees in its vicinity. It was planted by John Quincy Adams some time between the years 1825 and 1829, the years in which he was President. Although John Quincy Adams was the son of

to [illegible] varieties those that I find [illegible] propose to commence a nursery the next autumn and to plant acorns, hickory-nuts and chestnuts. But what is the season for planting, and how is the nursery to be managed? I should have commenced this process at least thirty years since, but I never had a permanent residence, and now I shall plant, if at all, more for the public than for myself."

John Quincy Adams did start a nursery. Scattered through his diary in the midst of lengthy accounts of affairs of state are notes like the following: "April 26, 1828. After the

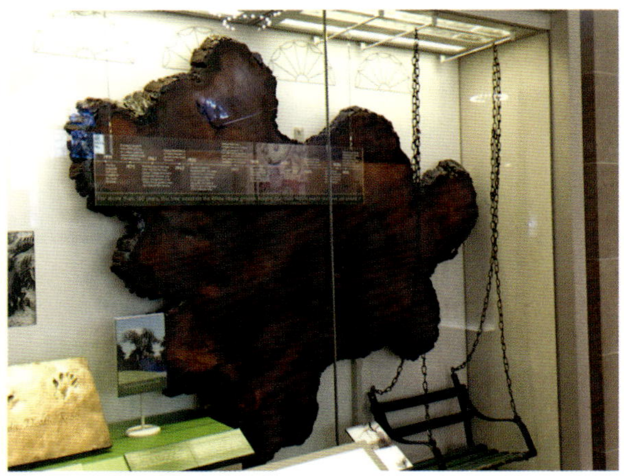

An analysis of this American elm, which grew on the South Grounds for more than one hundred years before it died in 1991, showed that it was planted during the John Quincy Adams administration. An article in the Evening Star *in 1921 celebrated Adams's contributions to the history of White House tree planting. The historic tree was captured in 1973 in this photograph (opposite) before it died. A section of the tree was preserved and is now displayed at the White House Visitor Center (above) with a timeline that tells the story of major historical events that occured during the life of the tree from 1827 to 1991.*

Adams also collected trees from other sources. He was pleased to acquire five different species of oak from the forests that still dominated much of the District of Columbia. He planted four rows of large walnuts sent from Pennsylvania and nine such nuts taken from the Adams family farm. Benjamin Crowninshield and Nathaniel Silsbee, both members of the House of Representatives from his native Massachusetts, presented him with English oak acorns from the city of Salem that may have come from the infamous gallows tree used to hang the condemned during the Salem Witch Trials of the 1690s.

Adams was almost a century ahead of his time in collecting the scions of these historic trees. A group called American Forests first started seeking out trees connected with historic persons, events, and places in 1917. By the late 1990s, people could buy seeds and seedlings descended from Johnny Appleseed's last known apple tree, a pin oak from Elvis Presley's Graceland estate, and even sycamore seeds that flew to the moon and back aboard Apollo 14. John Quincy Adams was a botanical trendsetter, a fact that no doubt would have shocked him at the time.

The Russo-American Oak

ON APRIL 6, 1904, Theodore Roosevelt planted a swamp white oak sapling with a remarkable heritage along the west side of the White House's South Portico. This tree's "grandfather" was a towering giant that had shaded George Washington's grave at Mount Vernon since the early nineteenth century. Sometime during Senator Charles Sumner's government career (1851–74), he sent several acorns taken from the tree to the czar of Russia, one of which is said to have been planted on an island outside the Imperial Palace of Peterhof. This descendant tree grew to maturity, and in 1898 Ethan A. Hitchcock, U.S. ambassador to the Court of St. Petersburg, determined to seek it out. A small tablet at its base announced: "The acorn planted here was taken from an oak which shades the tomb of the celebrated and never-to-be-forgotten Washington; is presented to His Imperial Majesty the Emperor of all the Russians, as a sign of the greatest respect."

The following year, Hitchcock was recalled to serve as the secretary of the interior. He returned to the United States with acorns from the Russian tree, which he then planted outside his home. By the spring of 1904, two young trees had grown from these seeds. One joined its patriarch outside Washington's tomb; the other was the tree that President Roosevelt planted. Fittingly dubbed the Russo-American Oak, this third-generation globe trekker flourished on the White House Grounds for several decades.

Franklin D. Roosevelt and Thomas Jefferson

EARLY IN HIS TERM, Franklin Delano Roosevelt suggested one of the most ambitious commemorative tree–planting programs the White House has ever seen: an entire grove dedicated to the memory of Thomas Jefferson, the president whom Roosevelt admired most. The idea was particularly poignant because Jefferson had never been able to plant anything on the President's Grounds in his own time. Despite his general push to thin and limit future tree plantings, the landscape architect Frederick Law Olmsted Jr. actually supported Roosevelt's proposal. He agreed that the grove would be an excellent way of honoring the tree-loving Jefferson, especially as it would physically and symbolically link the White House with the Jefferson Memorial, then in its early planning stages. Furthermore, Olmsted had the perfect place already picked out, a flat plateau near the eastern boundary of the far South Grounds that was sparsely treed and seemed to be lacking a purpose. A little later, Olmsted added a second location west of the South Fountain. In April 1939, the east and west groves were each planted with ten tulip poplars.

Roosevelt felt that the Jefferson trees should be native to the United States, and tulip poplar grows particularly well in Jefferson's home state of Virginia. Olmsted deemed the trees attractive, durable, and "highly characteristic of the region surrounding Washington." The groves greatly enhanced the appearance of the overall landscape, and filled in some unsightly gaps in the tree canopy. Roosevelt was extremely pleased with the result. Decades later Robert Redmond, a White House gardener from 1923 to 1963, recalled how the president personally inspected the freshly planted trees in a jeep. Sadly, none of the twenty tulip poplars survive today; they may have been damaged or destroyed in the course of the Truman renovation, 1948–52.

A tulip poplar is transplanted in 1952 following the Truman renovation of the White House. None of the poplars planted during the Franklin Roosevelt administration survive today.

The Jackson Magnolias

THE TWO MOST FAMOUS COMMEMORATIVE TREES on the White House Grounds, a pair of stately southern magnolias immediately west of the South Portico, are likely rooted more in myth than fact, but what a wonderful myth it is! The story begins with the presidential election of 1828, in which the incumbent John Quincy Adams sought reelection over his fiercest political foe, Andrew Jackson. Both sides savagely attacked one another through the press, mostly ignoring the issues and engaging in mudslinging and character assassination to such a degree that modern elections seem tame by comparison. Jackson emerged victorious, but not before his dear wife Rachel became embroiled in the torrent.

A flower from one of the two famous Jackson magnolias, which bears a marker (left) identifying the specimen as a southern magnolia planted by Andrew Jackson c. 1829.

The Adams camp branded her an adulteress for having married Jackson while still legally wed to another man. Rachel died suddenly on December 22, 1828, a mere two weeks after the newspapers announced her husband's victory. An anguished Jackson made it clear that he believed the political attacks against her good name had literally killed her, and he never forgave Adams. Jackson laid Rachel's body to rest on the grounds of The Hermitage, their Tennessee estate, on Christmas Eve. She was dressed in the white gown she had intended to wear to his presidential inauguration.

In the late nineteenth century a touching legend began circulating around Washington that told of Andrew's devotion to his wife even after her death. According to the story, the president brought with him to the White House seeds taken from one of the many southern magnolia trees growing on the grounds of The Hermitage. In life Rachel had loved these trees, and in death her husband would memorialize her by planting them on the President's Grounds. Doing so meant that Rachel would be with him in spirit, and he would have a tangible link back to the happy life they

This daguerreotype, made by John Plumbe Jr. in 1846, is the first photographic record of the South Front of the White House. It provides evidence of ornamental trees and shrubs flanking the South Portico, but despite the famous legend, the Jackson magnolia cannot be identified in this image. A recent view of the South Front of the White House (right) shows two mature magnolias on either side of the portico.

shared. Before long, this heartfelt tale was used to explain the origins of the two large southern magnolia trees growing immediately west of the South Portico, both of which pre-date White House horticultural records. The trees are not mentioned, however, in any accounts, not in government records or the popular press, dating back to Jackson's time.

Moreover, the Jackson Magnolias are not present in any drawing or painting until the latter half of the nineteenth century. They also seem to be absent in the 1846 daguerreotype taken by John Plumbe Jr., the first known photograph of the White House. This image captures a winter scene, as there is a patch of snow in the foreground and the deciduous trees are missing their leaves. Because the southern magnolia is evergreen (but not coniferous), Jackson's Magnolias should be readily apparent in the daguerreotype, but the only evergreen trees west of the portico appear to be either firs or spruces. There is also a row of some five or six evergreen shrubs running west and out of view to the left, likely yew or a similar species, but not young southern magnolias.

Given their absence from written and photographic written records, it seems unlikely that Andrew Jackson planted the magnolias during his presidency. This realization, however, does not in any way detract from the love story they have come to symbolize. That the legend has endured for more than a century is enough to ensure that the Jackson Magnolias remain the most historically significant trees on the White House Grounds.

Four saucer magnolias planted in 1962 during the Kennedy administration survive today (opposite) in the Rose Garden, just outside the Oval Office. President Kennedy's administration was tragically cut short before Kennedy had the opportunity to plant a tree on the White House grounds. These saucer magnolias serve as his commemorative trees.

During the Lyndon Johnson administration the East Garden was rededicated to Jacqueline Kennedy. Rachel Lambert Mellon's design for the garden is still in place today (opposite). As part of her work, Mrs. Mellon, seen above consulting with White House usher J. B. West in 1962, also planted this hearty orange, which survives today.

In 1964, President Lyndon Johnson, above with Lady Bird Johnson and Superintendent of Grounds Irvin Williams, planted a willow oak just south of the Oval Office. The mature tree continues to thrive today (right).

The cedar of Lebanon planted by President Jimmy Carter in April 1978 continues to thrive on South Grounds (right). When Carter planted the tree, he said, "This afternoon we are participating in a ceremony that has both enjoyable connotations and, I think, very strong historical and symbolic significance. In the Bible, the cedars of Lebanon are mentioned more than 60 times, and there are on the hills near Beirut trees that have been living there for more than 2,500 years, symbols of beauty and strength of an ancient and proud heritage. . . . I look on Lebanon with a heart filled with gratitude and appreciation, friendship, and also deep sorrow and concern."

The willow oak planted by President Ronald Reagan on Arbor Day 1988 continues to thrive today on the South Grounds.

An elm planted by First Lady Barbara Bush in 1991, seen opposite in 2011, no longer survives. It was propagated from an elm planted on the site by John Quincy Adams in 1826, which died in 1991. The two Japanese maples, seen in the lower left, planted by First Ladies Rosalynn Carter and Frances Folsom Cleveland, continue to grow on the South Grounds.

President George H. W. Bush is joined by Queen Elizabeth II with Prince Philip by her side in planting a little-leaf linden in 1991. The mature tree is seen opposite (in foreground) in 2011. The tree was planted to replace another specimen planted in 1937 by her father, King George VI, which was lost in a storm in September 1990.

President and Mrs. William J. Clinton plant a white dogwood on the South Grounds in 1996. The Clintons had also planted a white dogwood on the South Grounds in 1995 in honor of victims of the Oklahoma City bombing.

President Barack Obama plants a linden tree near the North Drive in 2009.

FOOD FOR THE PRESIDENT'S TABLE

IN THE TIME WHEN THE WHITE HOUSE was built, large estates were conceived as being self-sufficient, with gardens producing food for the household, and the President's House was no exception. Thomas Jefferson, the house's second resident, recognized the need for ancillary service buildings—stables, smokehouses, storehouses—and his terrace wings were the solution; they kept the functional outbuildings from cluttering the President's Grounds.

The Taft family brought a cow named Mooley Wooly to the White House, but she could not produce enough milk for the presidential family, so in 1910 Senator Isaac Stephenson of Wisconsin gave Helen Taft a registered Holstein named Pauline Wayne (seen above in front of the Eisenhower Executive Office Building). The cow arrived by train, weighed 1500 pounds, produced at least 8 gallons of milk a day, and bore a calf shortly after she arrived. The calf was named Big Bill (opposite). For three years Pauline Wayne grazed the President's Grounds, and when the Taft family left the White House she went back to Stephenson's farm in Wisconsin.

Jefferson's west wing had a "cow house," and throughout the nineteenth century dairy cows were often kept on the grounds to supply the first family with fresh milk, butter, and ice cream. Martha Patterson, President Andrew Johnson's daughter, who lived with her family in the White House, operated the dairy herself. Early each morning she would descend to the dairy, wearing a "calico dress and spotless apron," said an observer, to skim the milk for breakfast. The last presidential cow to graze on the President's Grounds was Pauline Wayne, and she was as much a pet of the William Howard Taft family as livestock. But if cows disappeared from the grounds in the twentieth century, vegetable gardens producing food for the president's table have been a recurrent feature, from Thomas Jefferson to Barack Obama.

PAULINE - THE PRESIDENT'S COW

The White House Kitchen Garden

WHITE HOUSE LORE CONTENDS that U.S. Capitol Architect William Thornton laid out a vegetable garden for President John Adams in the fall of 1800, shortly before Adams moved into the mansion. Vegetable gardens were then known as kitchen gardens, and Dr. Thornton is said to have established his on a patch of level ground northeast of the White House where Pennsylvania Avenue crosses the grounds. But when planting time arrived the following spring, Adams was back home in Massachusetts, and Thomas Jefferson had taken his place in the White House.

Bernard McMahon itemized the seeds he shipped to President Madison in 1809 with this bill of sale. He also included a short note apologizing for not being able to send the requested peas, cauliflower, and sassafras because of poor yields. He further stated that if Madison desired hyacinths, tulips, or other bulbous flower roots, he would be "happy to send him the strongest and best," and that he could also "furnish a very extensive assortment of the most ornamental and curious hardy perennial flowering plants."

Jefferson may or may not have planted a kitchen garden, but his landscape plan indicates he certainly intended to place one on the east side of the South Grounds. City Surveyor Nicholas King recorded that the area measured "two inches longer than the base, or freestone work," of the main block of the White House, or approximately 168 feet long. It would have been like his garden at Monticello, a mixture of ornamental and edible plants, set out in beds between gravel or grass paths. But the garden was not yet ready in the summer of 1806, when Jefferson turned down an offer from his gardening mentor Bernard McMahon for seeds and live plants. The president cited "the misfortune of my present situation as to things of this kind." The garden was still probably not ready when Jefferson left office in 1809.

During James Madison's first summer as president, a small package arrived at the White House labeled "The President of the United States (Garden Seeds)." The box contained a variety of vegetable seeds: cabbage, broccoli, savoy, radish, endive, cucumber, carrot, beet, parsnip, turnip, and leek. The accompanying bill of sale, totaling $14.12½ and dated June 27, 1809, provides a complete list of its contents. The parcel came from McMahon, and it is possible that Jefferson either placed the order before leaving office or did so afterward on Madison's behalf. It is certain that, at last, the president had a kitchen garden planted by the first White House gardener, Thomas McGrath.

There is no further mention of the White House garden for more than fifteen years, although it was evidently replanted after the 1814 fire and received more plants as time went on. By John Quincy Adams's presidency, it had fully evolved into the mixed-use garden that Jefferson intended. Writing in his diary on June 5, 1827, Adams described the scene in considerable detail:

> In this small garden, of less than two acres, there are forest and fruit-trees, shrubs, hedges, esculent vegetables, kitchen and medicinal herbs, hot-house plants, flowers, and weeds, to the amount, I conjecture, of at least one thousand. One-half of them perhaps are common weeds, most of which have none but the botanical name. I ask the name

B. H. Latrobe Esqr. Philadelphia June 27th 1809

for the President U. S.

Bought of Bernd. McMahon

3 oz Early York Cabbage	$1	.12½
2 oz Early Battersea do		.75
2 oz Large late Battersia do		.50
1 oz Sugarloaf Cabbage		.37½
2 oz White Broccoli		1.
2 oz Purple do		1.
2 oz Green and Yellow Savoy		.50
1 oz Red pickling Cabbage		.50
½ lb. Early frame Radish		1.
½ lb. Salmon-coloured do		1.
1 lb. Black & White Winter Radish mixed		2.
2 oz Curled Endive		.50
2 oz Long prickly Cucumber		.75
3 oz Long orange Carrot		.37½
3 oz Long red Beet		.37½
3 oz Dutch parsnip		.37½
½ lb Norfolk Field Turnip		1.
1 oz London Leek		.25
Box 25/100 Freight 50/100		.75
	$14.	12½

> of every plant I see. Ousley, the gardener, knows almost all of them by their botanical names, but the numbers to be discriminated and recognized are baffling to the memory and confounding to the judgment. From the small patch where the medicinal herbs stand together I plucked this morning leaves of balm and hyssop, marjoram, mint, rue, sage, tansy, tarragon, and wormwood, one-half of which were known to me only by name, the tarragon not even by that.

Later passages reveal that the vegetables included "deep blood-colored beet," "white-flowered carrot," "yellow-flowered parsnip," "enormous-leaved horse-radish," and "tall and slender-stemmed Jerusalem artichokes." Adams's collection of herbs also expanded to include mustard, thyme, shepherd's purse, and plants of the genera *Anthemis* and *Althaea*. Strawberries and red and black currants were special summertime treats. John Ousley even won a prize for his hautbois (musk) strawberries at the second annual Washington Horticultural Exhibition. Described as a "rich specimen of a rare sort," they impressed the judges and undoubtedly graced the presidential table on many occasions.

Some of the produce growing in the garden may have come from far-flung corners of the globe. Adams had long been a proponent of introducing and naturalizing all manner of foreign trees, fruits, vegetables, and other useful plants capable of being advantageously grown in the United States. Now, as president, he determined to advance this objective. At his request, Secretary of the Treasury Richard Rush directed U.S. consuls abroad to procure potentially beneficial plants unknown in America, "as well as superior varieties of such as are already cultivated here." Rush then specified that all "forest trees useful as timber; grain of any description; fruit trees; vegetables for the table; esculent roots; and, in short, plants of whatever nature whether useful as food for man or the domestic animals, or for purposes connected with manufacturers of any of the useful arts, fall within the scope of the plan proposed." Several pages of instructions followed, giving specific directions on how to properly pack various seeds, roots, and live plants for the voyage to America and requesting information on each plant's primary use, insect and frost resistance, and ideal growing conditions. Adams also called on the nation's ship captains to procure plants, and though it was difficult to keep seedlings alive on the open ocean, Adams received his first shipment three months later. More deliveries followed.

Shortly after becoming president in 1829, Andrew Jackson did away with the old mixed-use garden. He retained the approximately 2 acre plot southeast of the White House as an ornamental or flower garden and established a separate kitchen garden in Adams's former tree nursery on the southwest side of the grounds. As if to underscore this spatial division, James Maher, the public gardener who oversaw the development of all federally owned green spaces in Washington, hired a friend, William Whelan, to tend the new kitchen garden. Ousely remained in charge of the ornamental garden and, given their unequal rates of pay, was probably Whelan's supervisor as well. Ousley earned an annual salary of $450, while Whelan earned 87½¢ per day, the same wage as an unskilled laborer, and only for the days he actually worked.

The usual government records do not name the fruits, vegetables, and herbs that Whelan raised because he and Ousley did not purchase stock and seed. Instead they set aside a percentage of each year's harvest to be planted again the following spring. In those days gardeners and nurserymen also frequently traded their stocks, thus acquiring new plants for little or no cost. The only receipt for a purchase was a single order for four hundred asparagus roots placed by Ousley in 1849.

A surprising source for information about the president's garden is a speech by Whig Congressman Charles Ogle of Pennsylvania. On April 14, 1840, he rose to deliver a prepared address to the entire House of Representatives, which had gathered as a committee of the whole for the occasion. For three full days Ogle railed against President Martin Van Buren, a member of the rival Democratic Party and a candidate for reelection that year. Ogle accused the president of "spending the money of the People with a lavish hand" and living like a prince at the White House while the nation was in economic depression. Although Ogle titled his speech "The Regal Splendor of the President's Palace," it is known as the "Gold Spoon Oration" because he famously cited Van Buren's use of gold spoons as one of the many perceived presidential excesses.

Most of Ogle's diatribe focused on the opulence of the mansion itself, but he did not overlook the grounds. Ogle insinuated that the president was getting fat off the produce grown in the kitchen garden. To make his point, he listed fruits and vegetables by name: Neshanock potatoes, honest drumhead and early York cabbages, white and red sugar and pickle beets, marrowfat peas, carrots, parsnips, strawberries, dewberries, and raspberries. Ogle also identified "some exceedingly rare botanical and medicinal" herbs and was only too happy to call out the "vulgar" names of the most disconcerting sounding specimens: false fox glove, golden mad, golden club, enchanter's nightshade, dragon's head, lizard's tail, lion's foot, adder's tongue, monkey flower, virgin's bower, heart's ease, touch-me-not, and Balm of Gilead. The list might sound more like the contents of a medieval apothecary than the

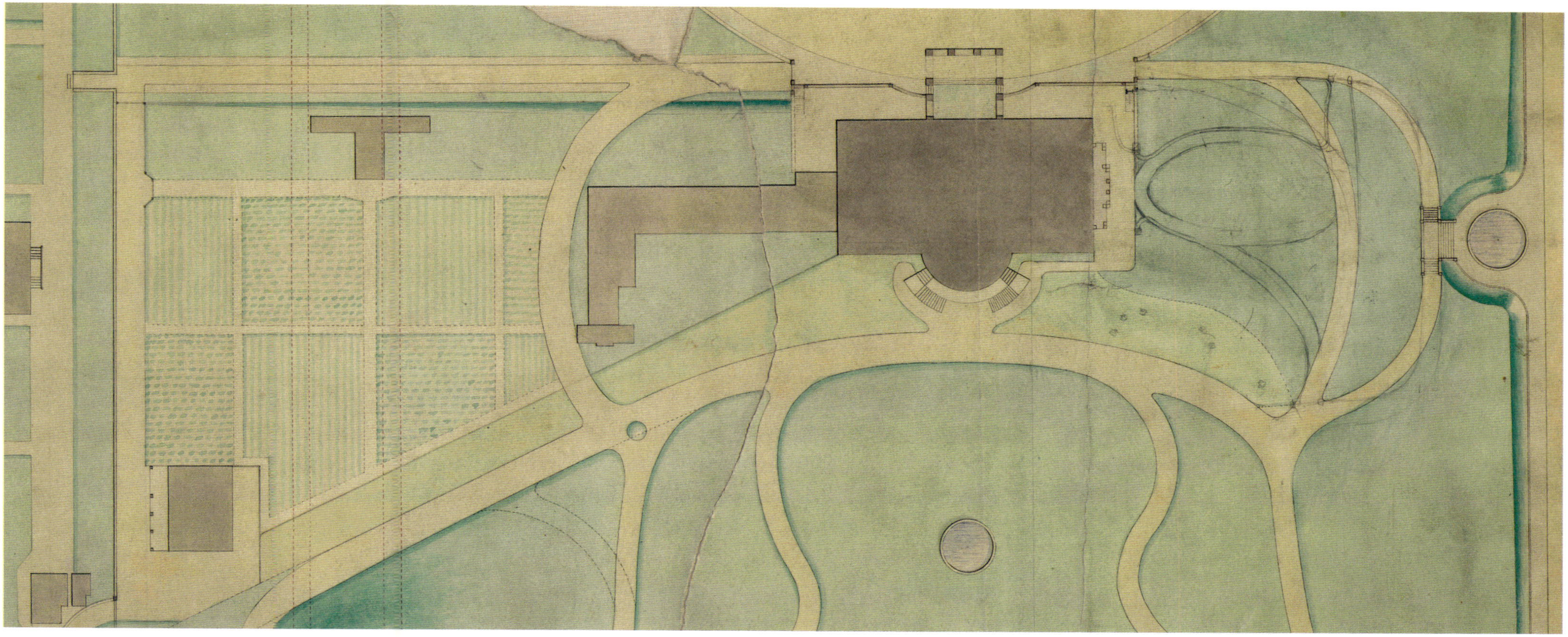

A landscape drawing of the White House Grounds done shortly after the close of the Civil War shows the kitchen garden as it likely appeared during Lincoln's presidency. It forms a trapezoidal shape and ranges over an area of almost an acre. The garden is also partitioned into eight separate sections, a fact that suggests it was run like a miniature farm.

garden of the president, but the false imagery suited Ogle's goal.

The next records about the White House kitchen garden date to the Buchanan administration. Between September 6, 1859, and June 19, 1860, John Watt, the White House head gardener after Ousley, purchased $66.66 worth of vegetable seeds and other growing stock, plus a few flowers and other decorative plants. The list is exhaustive and contains everything from spinach to pumpkin, onion to eggplant, and radish to rhubarb. Despite the huge size of this order, purchasing resumed again almost immediately and did not cease until June 1864, three years into the Lincoln administration. Adding all these new expenditures up yields a total of some $330 spent on plant stock for the kitchen garden alone in a span of slightly less than five years, no small sum during the mid-nineteenth century.

The government purchased all this material from a local man, John Saul, who owned a seed warehouse within the limits of the old City of Washington and ran a nursery farther out in the District of Columbia. His original bills of sale are cataloged in the Treasury Department vouchers preserved at the National Archives, and they provide the most complete record of the types of produce grown on the President's Grounds during historical times. The fact that this level of detail exists for the Lincoln years is welcome news. As with the seeds purchased for Madison in 1809, many of those included in Saul's 1864 list are still available today.

Not all the produce was destined for the Lincoln family's own use. First Lady Mary Lincoln made regular visits to Washington's military hospitals, carrying gifts from the White House gardens. A contemporary described these "errands of mercy" as undertaken to bring a small measure of relief to those who had given so much for the cause. Mrs. Lincoln went about "laying fresh flowers on their pillows and offering them delicacies brought from the White House. Her carriage would be laden with flowers and baskets of dainties, fruits, &c., for these hapless ones. In strawberry time her gardens would be stripped for their benefit."

The old White House kitchen garden remained in service for another decade and a half, until the construction of West Executive Avenue in the early 1870s triggered its removal.

John Saul's bills of sale for the Lincoln White House include selections such as lima beans, Brussels sprouts, corn, hubbard squash, kale, cayenne peppers, and raspberries, which remain popular today and are typically stocked at many garden centers and plant nurseries. Others, including "long green" cucumber, "Victoria" lettuce, "Winningstadt" cabbage, "refugee" beans, "Prince Albert" peas, and "orange" watermelon, are much less common now than in Lincoln's time but still available as heirloom varieties. The sheer number of choices is more than enough to keep the average backyard gardener busy for many years.

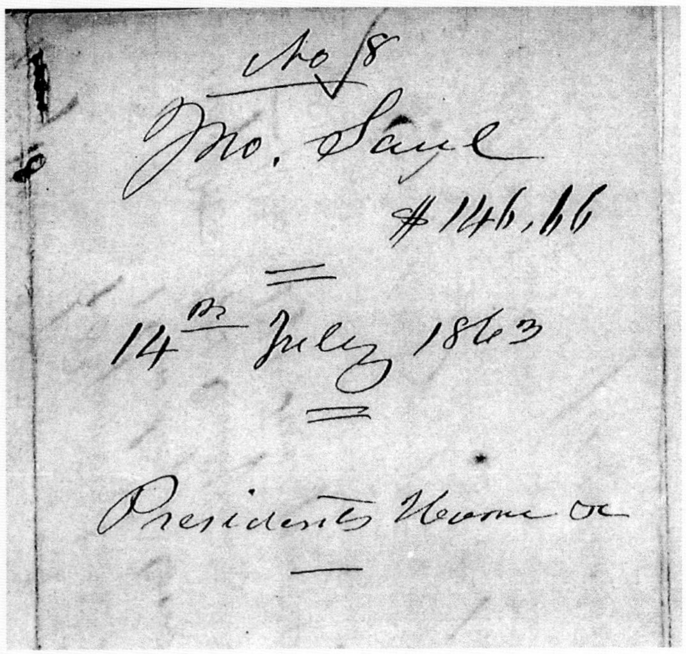

No 8

Jno. Saul

$146.66

14th July 1863

Presidents House &c

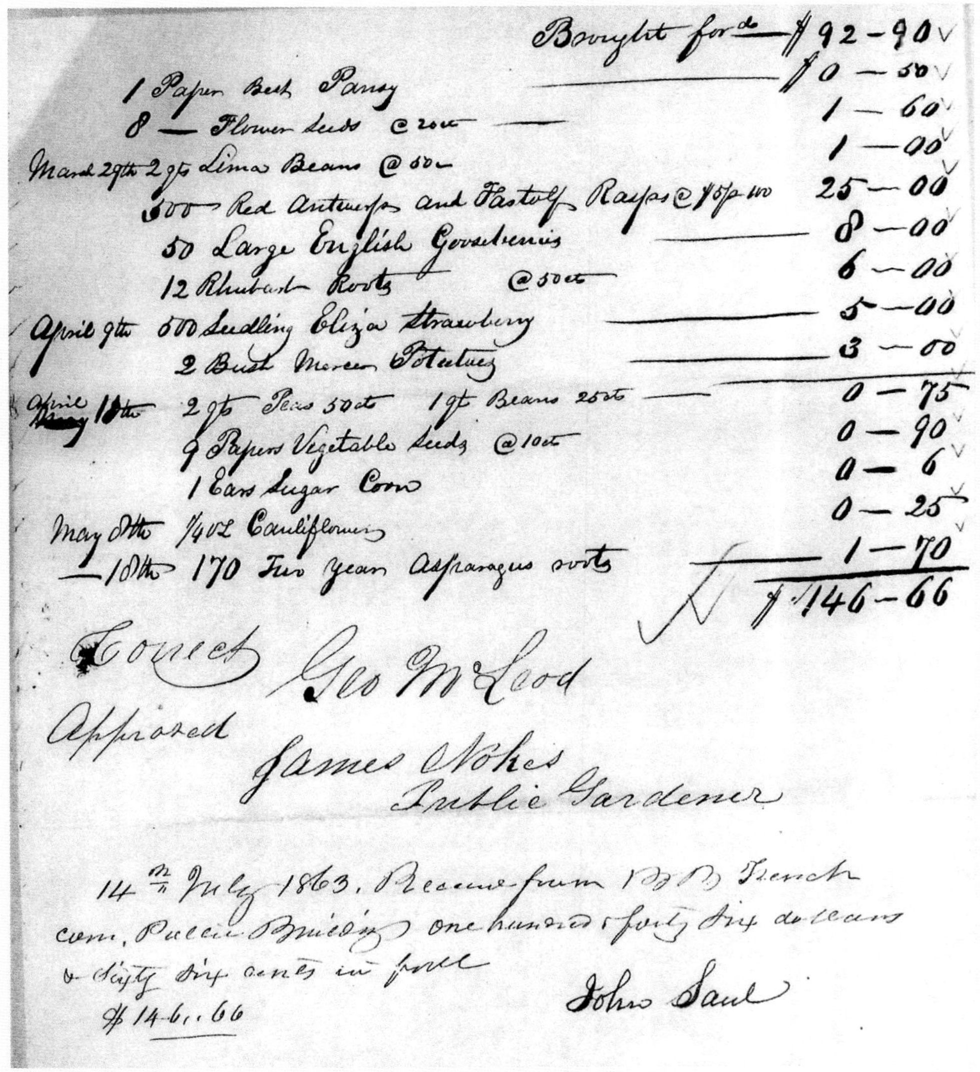

Brought for^d $92-90

1 Paper Best Pansy — $0-50

8 — Flower seeds @ 20cts — 1-60

March 29th 2 qts Lima Beans @ 50c — 1-00

500 Red Antwerps and Fastolf Rasps @ $5 p 100 — 25-00

50 Large English Gooseberries — 8-00

12 Rhubarb Roots @ 50cts — 6-00

April 9th 500 Seedling Eliza Strawberry — 5-00

2 Bush Mercer Potatoes — 3-00

April 18th 2 qts Peas 50cts 1 qt Beans 25cts — 0-75

9 Papers Vegetable Seeds @ 10cts — 0-90

1 Ears Sugar Corn — 0-6

May 8th 1/4 oz Cauliflower — 0-25

— 18th 170 Two year Asparagus roots — 1-70

$146-66

Correct Geo McLeod

Approved James Nokes Public Gardener

14th July 1863. Received from B B French com. Public Buildings one hundred & forty six dollars & sixty six cents in full

$146.66

John Saul

WASHINGTON, D.C., July 1st 1864

Commissioner of Public Buildings

Bought of JOHN SAUL,

NURSERYMAN AND SEED GROWER,

SEED WAREHOUSE, No. 396 SEVENTH STREET WEST, CORNER OF H STREET,
NURSERY ON SEVENTH STREET ROAD, 1½ MILES FROM BOUNDARY STREET.

FRUIT AND SHADE TREES, EVERGREENS, &C. | ROSES, DAHLIAS, PHLOXES, GLADIOLUS, AND DUTCH BULBS. | GENUINE GARDEN AND AGRICULTURAL SEEDS. | BEAUTIFUL AND RARE FLOWER SEEDS.

1863 Augt 26 1 lb Kale — 1 00
Sept 15 2 oz Lettuce Sorts — 1 00
1864 Jan 22 1 – Tomatoes 30¢ ½ oz Egg Plant 38 — 68
2 – Radish 25¢ 1 – Eg Cucumber 25¢ — 50
Feb 1 1 – Large York Cab — 30
13 8 qts Onion Sets @ 50¢ $4.00 ¼ lb Long Scarlet Radish 30¢ — 4 30
4 oz Flat Dutch Cab $1.00 1 oz Cayenne Pep 75¢ — 1 75
1 – Cherry Pep 75¢ 1 – Bull Nose 75¢ — 1 50
1 – Squash do 75¢ 1 lb Silver Skinned Onion $5.00 — 5 75
1 lb Blood Turnip Beet $1-50 ½ – Extra Ey Beet 75¢ — 2 25
4 oz White Solid Celery $1.00 2 oz Orang W. Melon 80¢ — 1 80
2 – G. Curled Savoy 60¢ 2 – Globe Savoy 60¢ — 1 20
2 qts Prince Albert Peas 70¢ 4 qts Champion Peas $2.00 — 2 70
2 – Queen of dwarf $1-20 ¼ lb R. T. Radish 37 — 1 57
¼ lb Whitetrip Rad 30¢ ½ – Curled Cress 50¢ — 80
½ oz Lesters Tomatoe 25¢ 2 – Purple top Turnip $2.00 — 2 25
1 lb Six week Turnip $1.00 ¼ – Gray Summer Rad 38 — 1 38
2 oz Hunters Cantelope 60¢ 2 oz Cluster Cucumber 40 — 1 00
8 – Long Green Cucumber $1-25 2 – Field Pumpkin 10¢ — 1 35
1 – Cashaw Pumpkin 20¢ 2. M. S. Water Melon 35 — 55
1 lb Okra $2.00 1 oz Morning Glories 37¢ — 2 37
1 oz Sweet Peas 38¢ ½ – Red Cypress Vine 75 — 1 13
1 – Leek 30¢ 1 lb Salsafy $4 — 4 30
1 lb Long Blood Beet $1-50 2 oz L. York Cab 60 — 2 10
2 oz Bunch Of Heart Cab 60 – 4 – L. Late Drumhead $1.00 — 1 60
4 – P. Flat Dutch do $1.00 2 – Late F. Dutch 60 — 1 60
1 – Drumhead Lettuce 30 – 1 oz Mignonette 60 — 90
1 – L. Mignonette 80¢ 1 – Gherkin Cucu 50 — 1 30
2 – S. C. Neck Squash 25 – 2 – W. C. Neck do 25 — 50
4 – Patty pan do 37 – 1 qt Refugee Beans 35 — 72
12 Papers Herbs $1-50 4 – Valentine Beans $1-80 — 3 30
14 Ears Corn assorted 81¢ ½ lb Ey Horn Carrot $1.00 — 1 81
½ lb Altringham Carrot $1.00 1 lb L. Orange Carrot $2 — 3 00
½ – Surrey do $1.00 ½ – S. do do $1.00 — 2 00
1 – Mangel Wurtzel $1.00 2 – Blood Beet $3.00 — 4 00
4 oz Flat Dutch Cab $1.00 1 lb Salsafy $4.00 — 5 00
1 lb Purple top Turnip $1.00 1 Paper Best Pansy 50¢ — 1 50
2 qts Lima Beans — 1 20

2 – G. Curled Savoy 60¢ 2 – Globe Savoy 60¢ — 1 20
2 qts Prince Albert Peas 70¢ 4 qts Champion Peas $2.00 — 2 70
2 – Queen of dwarf $1-20 ¼ lb R. T. Radish 37 — 1 57
¼ lb Whitetrip Rad 30¢ ½ – Curled Cress 50¢ — 80
½ oz Lesters Tomatoe 25¢ 2 – Purple top Turnip $2.00 — 2 25
1 lb Six week Turnip $1.00 ¼ – Gray Summer Rad 38 — 1 38
2 oz Hunters Cantelope 60¢ 2 oz Cluster Cucumber 40 — 1 00
8 – Long Green Cucumber $1-25 2 – Field Pumpkin 10¢ — 1 35
1 – Cashaw Pumpkin 20¢ 2. M. S. Water Melon 35 — 55
1 lb Okra $2.00 1 oz Morning Glories 37¢ — 2 37
1 oz Sweet Peas 38¢ ½ – Red Cypress Vine 75 — 1 13
1 – Leek 30¢ 1 lb Salsafy $4 — 4 30
1 lb Long Blood Beet $1-50 2 oz L. York Cab 60 — 2 10
2 oz Bunch Of Heart Cab 60 – 4 – L. Late Drumhead $1.00 — 1 60
4 – P. Flat Dutch do $1.00 2 – Late F. Dutch 60 — 1 60
1 – Drumhead Lettuce 30 – 1 oz Mignonette 60 — 90
1 – L. Mignonette 80¢ 1 – Gherkin Cucu 50 — 1 30
2 – S. C. Neck Squash 25 – 2 – W. C. Neck do 25 — 50
4 – Patty pan do 37 – 1 qt Refugee Beans 35 — 72
12 Papers Herbs $1-50 4 – Valentine Beans $1-80 — 3 30
14 Ears Corn assorted 81¢ ½ lb Ey Horn Carrot $1.00 — 1 81
½ lb Altringham Carrot $1.00 1 lb L. Orange Carrot $2 — 3 00
½ – Surrey do $1.00 ½ – S. do do $1.00 — 2 00
1 – Mangel Wurtzel $1.00 2 – Blood Beet $3.00 — 4 00
4 oz Flat Dutch Cab $1.00 1 lb Salsafy $4.00 — 5 00
1 lb Purple top Turnip $1.00 1 Paper Best Pansy 50¢ — 1 50
2 qts Lima Beans — 1 20

March 5 4 Bush Mercer Potatoes @ $2 — 8 00
— 30 2 doz Rhubarb Root @ $4 — 8 00
1 – Pears dwarf @ 50¢ — 6 00
2 doz Triomphe de Gand Straw $2.00 1 doz Grounden $1-50 — 3 50
1 lb Parsnip $1-50 ¼ lb Parsley 62 — 2 12
3 qts Early Peas @ 35 — 1 05
24 Pick New Roses — 15 90
April 22 1 oz Victoria Lettuce — 30
June 14 2 doz Choice New Dahlias @ $4 p doz — 8 00

$124 99

Correct
Geo McLeod,

Approved James Nokes, Public Gardener

10 August 1864
Received payment
John Saul

WITH THE LOSS OF THE KITCHEN GARDEN, no vegetables were grown on the President's Grounds until World War II, when a Victory Garden was installed south of the East Garden in 1943. After the war it reverted to grass. More than thirty years later, at President Jimmy Carter's request, an herb garden was planted for the first family's personal use. Fresh produce, however, was not harvested again until the Obama administration.

In 2009, First Lady Michelle Obama created a Kitchen Garden on the South Grounds to help educate schoolchildren on healthy eating. Children help Mrs. Obama plant and harvest seasonal vegetables and herbs to be used in the White House kitchen.

The beehive erected near the Kitchen Garden by the Obamas in the spring of 2009 yielded 134 pounds or roughly 34 gallons of honey in its first year. In 2011, President Obama purchased a home-brewing kit and by the following year had brewed three different beers using honey from the beehive: White House Honey Brown Ale, Honey Porter, and Honey Blonde.

On March 20, 2009, in the first spring of Barack Obama's administration, First Lady Michelle Obama and a group of local schoolchildren broke ground on a new 1,100-square-foot, L-shaped Kitchen Garden on the South Grounds near the White House fence line on E Street as part of an effort to initiate a national conversation on the health and wellbeing of our country. That conversation led to *Let's Move!,* an initiative Mrs. Obama launched in 2010 to help kids and families lead healthier lives. Throughout the Obama administration, Mrs. Obama invited kids from the Washington, D.C., area and around the country to help plant and harvest the garden. As a nod to history, heirloom vegetables favored by former presidents are often planted, including the "tennis ball" lettuce served as White House fare in Thomas Jefferson's time. The garden now spans 1,700 square feet, and the more than fifty varieties of fruits, vegetables, and herbs grown here provide food for the Obama family's table and are also incorporated into meals at White House events, including State Dinners. A portion of the produce also goes to Miriam's Kitchen, a local organization that prepares fresh, healthy meals for those in need. The Obamas also added a compost system and a honeybee hive, the first of its kind at the White House and now home to approximately 70,000 bees. In 2014, Mrs. Obama planted a pollinator garden next to the Kitchen Garden to support bees, butterflies, birds, and bats as part of the administration's efforts to promote pollinator health. The White House Kitchen Garden is a living legacy that inspires kids and families to eat healthy and reminds us of our agricultural heritage.

The Presidential
Garden Project
2015
REBUILD

ORNAMENTAL GARDENING

JOHN ADAMS WAS IN THE WHITE HOUSE much too brief a time to think of ornamental plantings. He was more a farmer than a planter, but, like his predecessors and successors from Virginia, he appreciated ornamental gardens, and after his presidency he installed some ornamental plantings around his farmhouse in Quincy, Massachusetts. Thomas Jefferson's love of gardens is well known, but he spent most of his presidency clearing away construction debris on the President's Grounds, filling in the pock-marked landscape, and, consequently, seems not to have overseen any ornamental plantings. Then came the War of 1812, the burning of the White House, and its reconstruction. Not until the time of John Quincy Adams was a proper flower garden in evidence on the President's Grounds.

The First White House Flower Garden

IN JOHN QUINCY ADAM'S TIME the mixed-use garden southeast of the White House bloomed with a variety of old-fashioned flowers, among them lilac, poppy, bladder senna, jonquil, periwinkle, archangel plant, wallflower, primrose, and cress. His diary entries also record the presence of shrubs and hedges in the garden, which were most likely used to line walks, complement other plantings, and screen views. Boxwood, a hardy, slow-growing evergreen shrub, probably made its first appearance in the garden collection around this time, and in subsequent years "boxwood edging" became an important part of its planting arrangement.

Colorful summer flowers line the walkway in beds outside the East Colonnade continuing a long tradition of ornamental gardening on the White House Grounds.

The next president, Andrew Jackson, ordered the remarkable plot separated into a kitchen vegetable garden and an ornamental flower garden. The former moved southwest of the White House and replaced the tree nursery established by Adams, while the latter remained in place to the southeast. The flower garden was formal like those of the English colonial era, with flowerbeds lined by curving walkways, for promenading. The garden was surrounded by a wooden paling fence, and a surviving Treasury voucher records that a local blacksmith named Caleb Buckingham fashioned a small but attractive iron entry gate.

Flower seeds and young dahlias for Jackson's garden were purchased, the William Prince & Sons Nursery of Flushing, New York, and more than 1,400 "bulbous flower roots" from Edward Dyer's Washington, D.C., auction house. Although Dyer's vouchers do not identify these plants by name, the large acquisition likely included tulips, daffodils, crocuses, irises, hyacinths, lilies, and narcissus, all common flowers grown from bulbs. Since Dyer was a local businessman, these bulbs likely came from a local source, and in the 1830s there was only one major grower operating in the area, William Pierce's Linnaean Hill Nursery. Since its opening in the 1820s, Linnaean Hill had been perhaps the leading supplier of fruit and forest trees, flowers, shrubs, and other plants used in landscaping the city's streets, estates, and public places. Head Gardener John Ousley no doubt procured other orna-

mental plants by trading seeds and cuttings. Before Jackson left office a local construction firm supplied wooden settees and trellises, several of which were specified as arbors and espalier frames for rosebushes and climbing vines. Later, a 140-foot-long grapevine trellis was also erected somewhere on the South Grounds, possibly in the ornamental garden.

Jackson's successor, Martin Van Buren, first developed an appreciation for ornamental gardening while serving as U.S. minister to Great Britain in the early 1830s. During this time he toured England alongside American author Washington Irving, taking notes on the country houses and their gardens. The high point came when King William IV personally received him at Windsor Castle and showed him its extensive grounds. Van Buren, in turn, promoted the development of the gardens and grounds at the White House. During his "Gold Spoon Oration" Congressman Ogle decried this connection and described the ornamental garden as corresponding "in its general arrangements with the style and fashion of some of the most celebrated royal gardens in England. It has a choice collection of both native plants and exotics, many of the latter having been gathered from almost every clime." He specifically mentioned "numerous families" of tulips, lilies, pinks, and roses, as well as honeysuckle, trumpet creeper, primrose, and lady slipper and alluded to "many thousand other sweet flowers and shrubs, which all the lovers of beneficent Nature admire." By this Ogle did not intend a compliment but rather disdain for what he perceived as an unnecessarily lavish lifestyle at the White House.

Apart from Ogle's speech, there is little record of the ornamental garden during Van Buren's tenure. Among the very few relevant documents is a payment of $74.25 for four dozen "large daily" rosebushes, two verbenas, and a petunia plant, made out to William Buist, who operated a small nursery three blocks from the White House, at the corner of Twelfth Street and Pennsylvania Avenue, and was likely a relative or Robert Buist, whose Philadelphia-based nursery was among the largest in the nation. Van Buren also purchased cast-iron settees and chairs for the flower garden, as well as circular benches that workmen installed around some of the larger trees on the President's Grounds.

The White House ornamental garden approached its full splendor in the late 1840s during the presidency of James Polk. It boasted a staggering number of flowers, and almost every month brought new plants into blossom. John Ousley planted some one hundred varieties of roses alone, including the moss rose, a small and favorite species in the nineteenth century.

By 1855, the ongoing expansion of the U.S. Treasury Building literally cast a long shadow over the ornamental garden. Although the garden had been a fixture on the landscape for three decades and was much loved by presidents and their families, the impending construction of Treasury's west wing now called its permanency into doubt. Back in 1838, Thomas U. Walter, the architect of the Capitol, had warned about encroachment onto the President's Grounds, recommending the relocation of all the executive department buildings, including the Treasury. But Robert Mills, the architect of the new Treasury Building, had assured critics that his grand neoclassical structure "was to be in harmony with the president's grounds when completed." Now, nearly two decades later, Walter had been proved right and Mills wrong: there was simply not enough room to accommodate both the White House ornamental garden and the completed Treasury Building.

On February 2, 1857, the Treasury Building's construction superintendent reported that he could go no further without impinging upon the garden. Government officials now scrambled for a solution. One was to relocate the garden elsewhere in the city, but then, of course, it would not be the president's garden. Another was to place it back alongside the kitchen garden, southwest of the mansion. But already the War and Navy Department Buildings, then located immediately west of the White House Grounds, were expected to expand. It was ultimately decided to continue the Treasury wing and remove the old ornamental garden entirely. As a concession, the White House would receive a conservatory, to be built atop the mansion's West Terrace.

During the second half of the nineteenth century, flowers and other decorative plantings on the White House grounds appeared only in ornamental beds placed here and there about the landscape, with the largest concentration along the border of the mansion's South Front and around the various fountains. In 1897, a total of seventy flowerbeds bore 63,205 spring bulbs alone. At this time the grounds were awash in flowers, among them hyacinths, tulips, crocuses, scillas, liliums, narcissus, pansies, myosotis, daisies, cannas, geraniums, achyranthes, coleus, althernantheras, echeverias, begonias, gladiolus, tuberoses, and lantannas. Most were planted by the thousands; some by the tens of thousands.

Thousands of tulips are planted throughout the White House grounds each year. The beds seen here are in the Rose Garden.

Andrew Jackson Downing's Fence

As late as the 1850s, flowers were concentrated in the ornamental garden, although ornamental trees and shrubs flanked the South Portico. Andrew Jackson Downing favored retaining and enhancing these embellished areas on both sides of the portico, which he jointly referred to as the "border around the President's house." Although he left no records hinting at its planned layout or composition, the inference is that he intended it to become a highly decorative space. The first flowerbeds appeared inside this area in the 1870s during the Grant administration. Downing personally selected the wire fence put up to separate this border from the rest of the South Grounds, and it became his only tangible addition to the White House landscape.

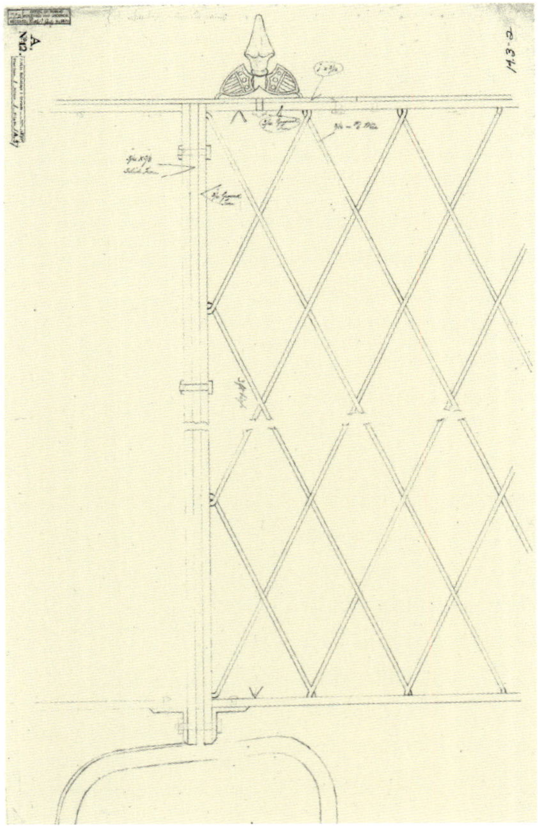

The National Archives houses a large pencil sketch that appears to reproduce the features of the ornamental ironwork fence purchased in 1851 with nearly life-size detail. It was obviously much lighter than the heavy iron perimeter fencing that encloses the White House Grounds, more of a latticework screen than a serious barrier. Topped by small finials bearing miniature U.S. Shields, this fence was not affixed to the ground but merely rested on short legs.

Downing's purpose for the fence was to politely deter encroachment, and he specifically chose a design that made this point unobtrusively. He noted that a "simple pattern not calculated to attract the eye" was far better for his purpose than a more substantial or "showy" one. Outside a private residence such a fence would have been unnecessary, but as the public had regular access to the President's Grounds, a fence was needed to protect the flowers and other ornamental plants that would border the mansion.

Purchased in the summer of 1851 from J. B. Wickersham, a manufacturer of ornamental ironwork in New York City, this fence edged the gravel carriageway that skirted the South Portico and traversed the lawn from east to west. It remained in place for some eighty years until, being rusted out and broken, it was finally removed during the first term of Franklin Roosevelt.

This photograph of c.1875 clearly shows Downing's fence and the first White House greenhouse (the relocated orangery) at the far left. The inconspicuous nature and, even more important, the ability to easily see through this fence, designed to protect the ornamental plants along the gravel carriageway, allowed the border to become a part of the larger landscape rather than a wholly sequestered place.

Gardening Under Glass: The First White House Greenhouse

THE DECISION TO CONSTRUCT a conservatory stands out as a seminal moment in the history of the presidential landscape. It was not, however, the first instance of gardening under glass at the White House. Thomas Jefferson tended a variety of potted plants, including roses, geraniums, strawberries, and figs, which he set on sunny windowsills in his office, today the State Dining Room. He placed glass-domed bell jars, or cloches, over the most delicate plants to keep in the warmth and shelter them from the drafts that permeated the recently completed mansion.

The first White House Conservatory, pictured here in the late 1860s, was large, covering the majority of the West Terrace. It featured a high glass ceiling and sides and was connected to the White House via a short glazed passage approximately 12 feet long.

Charles Bizet, White House gardener under James Monroe, took this idea a step further. He placed the first cold frames and hotbeds outdoors, probably in the garden area southeast of the house. A cold frame is essentially a bottomless box with a glass top placed over a small plant to protect it. As with a cloche, sunlight shines through the glass and warms the plant, allowing it to be kept outside even after the air temperature cools in the winter. A hotbed is a cold frame that also employs an artificial heat source, and in the early nineteenth century this added heat almost always came from fermenting manure. Additional hotbeds were added to the White House Grounds through at least the 1840s, and because of their fragile nature required regular repairs.

During the administration of John Quincy Adams, John Ousley needed a place to store his tools and supplies, and he quickly adapted the old fireproof Treasury vault, built as part of Thomas Jefferson's unrealized plan to connect the White House and executive buildings, to his purpose. It had stood empty and unused since the 1814 fire and was ideally located. Adams's cold-averse citrus trees, including lime and persimmon, may have overwintered inside the former vault.

President Jackson had the building converted to serve this purpose full time in 1835, and the old vault became an orangery. As with all orangeries, that at the White House featured a conventional tile roof; true glass-roofed greenhouses did not come into widespread use in the United States for another two decades. Instead, a full bank of large glass windows let in the southern sun, and heavy masonry walls trapped the heat. The White House orangery also had a stove to warm the interior during cold winter nights. Jackson may have ordered the creation of the orangery to provide a safe place for one of his most prized possessions: a Malayan sago, or feather, palm that had once belonged to George Washington. It had been offered to Jackson by Washington's heirs during the bitter winter of 1835, after the orangery at Mount Vernon burned.

The new orangery now afforded a place to grow citrus trees year-round without fear of frost. Lemons and, fittingly, oranges joined the limes and persimmons. The collection soon further expanded to include other temperate and tropical plants unlikely to survive if left exposed to cold weather.

Camellias bearing red, white, or pink blossoms were favorites. As a young girl Jessie Benton, the daughter of Senator Thomas Hart Benton and future wife of explorer John C. Frémont, recalled the East Room, decorated for one of Jackson's large receptions, as a "silent waiting fairyland" festooned with "stands of camellias and laurestina banked row upon row, the glossy dark green leaves bringing into relief their lovely wax-like flowers." Ousley borrowed the live plants from the orangery for the occasion, and given the description must have left it temporarily bare.

During the summer of 1853, Jackson's orangery became a true greenhouse through the addition of new wings with low-sloping glass-paned roofs. The new structure retained the orangery as its centerpiece and incorporated it as a private presidential solarium. John Watt, head gardener at that time, quickly filled the expanded space with more fruit trees and flowers, many of which came from John Saul's nursery just outside the city limits. Watt also procured star jasmines and a wide variety of camellias from the Philadelphia nursery of Peter MacKenzie, probably the largest supplier of camellias in the United States. In addition, Joshua Pierce, another local grower who had probably stocked the White House with plants since at least Jackson's time, developed a reputation as an expert camellia grower and likely filled the greenhouse with his stock. Another supplier provided azaleas, begonias, china pinks, sweet alyssum, still more camellias, and a variety of flowering shrubs and annuals. The camellias remained the stars of the greenhouse, and Watt selected particularly beautiful varieties, including The Queen of Whites, Lady Anne's Blush, Duchess d'Orleans, Palmer's Perfection, Miss Perceval, Douglass White, and Varigata.

Since the general public was not allowed inside the greenhouse and no architectural drawings of it have survived, carpentry, painting, and plumbing accounts provide the only information about its interior. The older, center section featured brick floors, holdovers from the orangery's former life as the fireproof vault for the first Treasury Building. In the new wings floors were of bare earth and covered with moist sand for the benefit of the fruit trees and other large plants that occupied the majority of the floor space. These plants were potted in square wooden boxes purchased from suppliers in Washington and Baltimore, some in standard sizes but others made-to-order for specific specimens. After the weather warmed in the spring, Watt and his assistants moved most of the tubbed plants out into the ornamental garden and returned them safely to the greenhouse before the first winter frost. Shelves and stepped risers filled with flowers and smaller potted plants ringed the rooms, which were all painted white and accented with green doors. A system of cords and pulleys opened and closed the ceiling sash, thereby adjusting the interior temperature and providing fresh air. Located out of sight in the cellar, an impressive coal-fired furnace heated the entire building, and a system of tin pipes similarly supplied running water from underground cisterns.

Over the next four years President Franklin and First Lady Jane Appleton Pierce made full use of the greenhouse, enjoying the privacy it afforded them and taking more and more cut flowers to display within the White House. Before the 1850s, wax flowers were most commonly used as decorations, owing to the prevailing belief that the pleasant smell of real flowers also contained dangerous "effluvia." While this misconception was not entirely discounted by the time Pierce took office, the use of cut flowers had become popular with the social elite across the United States. In addition to arrangements placed around the home in vases or on "stands," fashionable ladies quickly took to wearing cut flowers in their hair and in corsages. Most of the early cut flowers used at the White House were camellias. In her memoirs, Virginia Clay-Clopton, whose first husband was Alabama senator Clement C. Clay, recalled the "remarkable bouquets" placed at each seat for a State Dinner. Each bouquet contained at least six camellias "as big round as a breakfast plate," and guests could take them home wrapped in shells of "marvelously wrought lace-paper." Soon this custom began to be repeated at engagements throughout Washington. For an entire season, said Mrs. Clay, the camellia was "the only flower seen at the houses of the fashionable or mixing in the toilettes of the belles." So began the long-standing tradition of fresh floral arrangements at the White House.

Despite the recent renovation of the greenhouse and its mounting acclaim, the structure's continued existence was far from certain. Like the ornamental garden, it stood directly in the path of the expanding Treasury Building. As part of the decision to remove the garden and construct a conservatory for the president atop the West Terrace, the greenhouse was moved to the southwest side of the White House Grounds. In the fall of 1860, workmen dismantled the structure brick by brick and reassembled it at the far end of a north-south extension of the West Terrace originally built in 1820 as a stable for President James Monroe. Once in its new location, the greenhouse became a purely functional space, growing cut flowers for the rooms of the White House and live plants for display in the Conservatory, which was now an extension of the White House itself. An interior stair allowed the gardeners to move plants from the greenhouse to the Conservatory without having to take them outside, a convenience that proved particularly important during the winter, when cold air would have damaged or killed delicate specimens.

In this c. 1860 view of the Conservatory, George Washington's sago palm can be seen in the left foreground.

The Great Glasshouse Complex

THE NEW WHITE HOUSE CONSERVATORY followed a design by Edward Clark, the architect in charge of extensions at the old Patent Office and Botanic Garden. The entire process, from planning to completion, took only nine months. The finished structure stood atop the West Terrace by the end of October 1857, about halfway through President Buchanan's first year at the White House. With the weather staying unusually warm, the extensive collection of rare and valuable plants was safely transferred to this new glasshouse in November.

From the outset the Conservatory, connected to the White House by a 12-foot glazed passage, served as a private setting for the president and first family to enjoy and entertain small groups of select guests. Its location assured that the privacy and seclusion of the old ornamental garden transferred to this indoor retreat and also that the White House, notoriously dry and stifling during the winter months, benefited from its humidity.

Details about the Conservatory's interior arrangements come from prints and photographs. The main, central room overflowed with plants—in hanging baskets suspended from the ceiling, hundreds of terra-cotta flowerpots ranging over rows of heavy tables and shelves, larger tubs arranged on the floor, and vines reaching toward the ceiling. George Washington's sago palm retained a place of honor, and John Watt purchased more than $2,000 worth of additional plants from Peter MacKenzie's nursery. In addition to roses and camellias, by this time indispensable decorations at White House events, live heliotropes, violets, hyacinths, and other rare hothouse plants, all borrowed from the Conservatory, began appearing at important White House receptions.

Entitled Miss Lane's Conservatory, at the W.H., Washington, *this hand-colored engraving depicts the Conservatory as Harriet Lane, James Buchanan's niece and his White House hostess, would have known it in the late 1850s.*

Watt and his assistants utilized three hydrants, two in the main room and one in the adjacent potting shed, to keep the plants in the Conservatory well watered. The large furnace, which heated a system of hot-water radiators, was estimated to consume fifty tons of coal per year. Basins of water put more humidity into the air, and there might have been a small fountain as well. Clark advocated for a fumigation room and water tanks for the growth of aquatic plants, both of which might have been incorporated into the Conservatory's design at a later time.

The young and popular Harriet Lane, White House hostess for her bachelor uncle President James Buchanan, dearly loved the new Conservatory, and in 1860 she led a tour of it for Albert Edward, Prince of Wales, on a royal visit to the United States. A news reporter also managed to get inside and described the place as "almost like penetrating the luxurious fragrance of some South American island, so warm and odiferous is the atmosphere. . . . Here you may see orange trees, and a lemon tree . . . rows of prickly cactus of every size and shape . . . camellia japonica, delicate spirea, ardisia, and poinsettia."

First Lady Mary Lincoln also enjoyed the steamy Conservatory as one of the few parts of the White House and its grounds that remained a private space throughout the course of the Civil War. She quickly developed a close friendship with Head Gardener John Watt, who delighted her with a seemingly endless supply of fresh-cut flowers. Sophisticated arrangements decorated the presidential table during important dinners and poked out of antique vases formerly used to display wax flowers. Mrs. Lincoln also received a fresh bouquet each day, which she often sent on to one of the fashionable ladies of the city.

The Southern Plains delegation posed for photographer Mathew Brady in the White House Conservatory on March 27, 1863. The Native Americans in the front row are, left to right: War Bonnet, Standing in the Water, and Lean Bear of the Cheyennes, and Yellow Wolf of the Kiowas. The identities of the Native Americans of the second row are unknown. John Nicolay, Lincoln's secretary, is seen at the center of the back row. Within eighteen months from the date of this sitting, all four men in the front row were dead. Yellow Wolf died of pneumonia a few days after the picture was taken; War Bonnet and Standing in the Water died in the Sand Creek Massacre; and Lean Bear was killed by troops from Colorado Territory who mistook him for a hostile.

Mrs. Lincoln's great demand for cut flowers necessitated the regular purchase of live plants and seeds in previously unheard of quantities, despite the shrinking size of congressional appropriations. John Saul's nursery seems to have again been the principal source, and Watt's old associate sent along a wide variety of flowers: dahlias, morning glories, hollyhocks, larkspurs, asters, pansies, petunias, phlox, gillyflowers (Emperor and German stocks), bleeding hearts, hyacinths, tulips, crocuses, gladiolus, balsam, chrysanthemums, cineraria, Chinese primroses, and cypress vines. Watt also purchased hundreds of camellias, roses, and other greenhouse plants. Finally, an unknown, but evidently large, quantity of ornamental plants in the Conservatory were on loan from the recently widowed Sarah Lee Gales, whose husband Joseph was publisher of the *National Intelligencer* newspaper, an official printer to Congress, and a former mayor of Washington. Orders for thousands of terra-cotta flowerpots, ranging in size from 5 to 15 inches, attest to the vast scale of the enterprise.

At 8:15 a.m. on January 18, 1867, an alarm rang out at the White House: the Conservatory was on fire! Steam fire engines arrived almost immediately, but it took two hours to control the blaze, as it originated in the 3-foot gap between the floor of the glasshouse and the roof of the terrace, and the firemen could not pump water directly into this narrow space. Instead they concentrated on making sure the fire did not spread to the mansion. Officer Andrew C. Smith, the first policeman assigned to permanent duty at the White House, heroically saved the small greenhouse, the relocated orangery attached to the West Terrace, but he became "insensible from cold and suffocation, and was carried into the house apparently dead." He revived a few hours later, but remained in a "precarious condition" for several days. Structurally, the Conservatory was destroyed, and "at least one-third" of the plants died. The losses included the sago palm, brought to the house by Andrew Jackson, orignially from Washington's orangery at Mount Vernon.

The ruins were cleared away by the end of winter, and the West Terrace received a new roof in the spring. The Conservatory itself was rebuilt as it had been, with a wooden superstructure over the same 3-foot gap. Completed in 1869, it was just as susceptible to fire as its predecessor. An iron frame would have been more fireproof, but Congress declined to fund it, so ensuring that maintenance costs for the wooden building would recur. The Conservatory was fully restocked with plants by 1870, and the old greenhouse was expanded. Its new section housed a grapery with "a large number of the choicest vines." This addition stood on the west side of the greenhouse and resembled a lean-to with a steeply pitched roof angling down to the south.

An inspection in 1873 revealed that the new Conservatory was already in need of significant repairs, for the high temperature and humidity had proved ruinous to its wooden interior. The floor boards were so rotted that most could not be salvaged. A new wooden floor was installed, and under it a system of brick arches and concrete-covered metal beams. The old wooden staircase leading down to the greenhouse was also replaced with stone steps, and, despite the added strain on the rafters, a "hoisting machine" was added so gardeners could easily lift and move even the heaviest of specimens. A year later, Congress delivered a special $10,000 appropriation "for continuing the permanent improvement" of the Conservatory, but what it really needed was an iron superstructure that cost five times this amount. Instead the appropriation just funded the never-ending cycle of repair work.

New York.
Brady

For the presidents, however, the Conservatory remained as popular as ever. To meet the increasing desire for cut flowers, it was extended during the Grant administration to allow orchids to be grown in large numbers for the first time. As most orchids are tropical species, they were virtually unheard of in the United States until the mid-1800s. The few fragile plants that survived long sea voyages to America often died at the hands of amateur enthusiasts who knew little of orchid culture. However, growing techniques improved dramatically after the Civil War, and successful hybridization efforts resulted in not only a profusion of new orchid varieties but also plants that were more vigorous and less fatally delicate. The result was an "orchid fever" that swept through the social elite who could afford the high prices that the flowers commanded. No exception, First Lady Julia Grant directed the establishment of a prize-worthy orchid collection at the White House. Rather than relying solely on domestic nurseries, she followed the example of John Quincy Adams in requesting the State Department to instruct its consuls abroad to procure as many varieties as possible. Decades later she was credited with ensuring the orchid's widespread popularity. "When Mrs. Grant was presented with some blooms by a South American minister, the orchid, then practically unknown, was destined to become the aristocrat of all flowers," said a newspaper in 1920.

Julia Grant's love of orchids no doubt prompted the building of three additional greenhouses on the White House Grounds. They were long, narrow structures erected to the west of the White House and, like the Conservatory, roofed with glass; later they were identified as the geranium house, orchid house, and small rose house. They were in place by 1873, when the author and journalist Mary Clemmer Ames described the Conservatory's interior:

> Leaving the state dining-room behind, we pass through the western wing into the conservatory, one of the largest in the country. It is a favorite resort for lady and gentlemen promenaders on reception days, lined, as it is, on either side with the bloom and fragrance of rare exotics. A large aquarium stands at one end, and a short passage and flights of steps lead down to a greenhouse and grapery filled with flowers and luscious fruit. Three other greenhouses flourish in the gardens west of the mansion.

The new greenhouses were sorely needed. Even without the new orchid collection, the existing facilities were severely overtaxed by the demands of the White House, and outside orders were sometimes necessary to supplement the flowers and other decorations grown on-site.

In 1880 a new, larger rose house was added to the glasshouse complex, placed where the Rose Garden is today. The structure was built at grade with no underlying foundation, its bare dirt floor crisscrossed by paved walkways. Full south sun streamed in through the glass outer wall and roof, and rolled awnings could be lowered if shade was desired. Head Gardener Henry Pfister and his assistants operated the rose house, like the other greenhouses, as a functional rather than decorative space. The goal was to grow as many plants as possible.

The Conservatory was, by contrast, an aesthetic space, and before the end of the Hayes administration it was also extended to the west side of the White House proper, thus replacing the former glazed passage, which had most recently served as President Grant's Billiard Room. The addition was considerably wider and taller than the existing glasshouse already in place atop the West Terrace. Two of the White House's original second story windows were now interior to the enlarged Conservatory, opening into the State Dining Room. Moving north along what had once been the exterior face of the mansion's west wall, two large rockwork water pools alternated with these windows. Then came the Conservatory's entrance, an open passageway leading down two stone steps from a State Floor hallway and flanked by narrow windows punched out between sets of eighteenth-century Ionic pilasters. Given all the alterations to the historic building fabric, the process was a bit like forcing a square peg through a round hole, but President Rutherford Hayes and First Lady Lucy Hayes both wanted a larger Conservatory and would not be denied.

To enter the expanded Conservatory was to step into a tropical paradise. Just past the entrance was a large open area ringed by flowerbeds; beyond were two long, straight walks extending all the way to the far (west) end where they met one another, forming a semicircle. The resultant horseshoe-shaped route thus allowed one to cover the entire distance without backtracking. Beyond their own enjoyment, President and Mrs. Hayes found a novel use for this "Conservatory Promenade." Traditionally, liqueurs had been made available to White House guests following dinner, but no alcohol of any kind was served in the White House during Hayes's term. This prohibition originated with Mrs. Hayes, who disapproved of drinking on moral grounds and supported the temperance movement. Her husband saw nothing wrong with an occasional drink but decided that banning alcohol would help showcase the strong Christian values he had made a hallmark of

Orchids are seen the Conservatory in this 1889 photograph. An "orchid fever" swept the nation in the late nineteenth century after gardeners learned how to cultivate these fragile tropical plants. Ultimately an entire greenhouse was devoted to their care.

Instead of serving cordials and liqueurs after dinner, President and Mrs. Hayes led their guests on a "Conservatory Promenade," strolling the horseshoe-shaped path through flowers and ferns.

his presidential campaign. This abstinence policy annoyed some, but more than anything else it resulted in a rather awkward after-dinner situation, where coffee was the only beverage on the menu. Most guests, however, quickly forgot about the missing wine and cordials upon being invited on a tour of the Conservatory. After everyone finished eating, President and Mrs. Hayes would rise, join hands, and lead a procession that often numbered well over one hundred men and women into the glasshouse. As the wide-eyed guests stepped along on encaustic tiles past mounds of ferns and palms and row after row of flowers raised on stepped tables, the first lady pointed out rare plants and particularly lovely specimens. After completing the circuit, she was happy to chat with those who wished to linger amid the vivid colors and sweet smells.

Most of Mrs. Hayes's botanical knowledge came firsthand. A presidential aide later recalled that she "spent much time and thought in the conservatory, where she could often be seen at practical work among her plants, trowel in hand or pruning-knife, teaching Fanny and little Scott how to care for the beautiful buds that came into being, and matured, and gave way to others." She was in the Conservatory with her children one winter day in 1879 when a photographer took what is probably her most famous picture.

Shortly after moving into the White House for the first time in 1885, President Grover Cleveland began taking strolls in the Conservatory, admiring the flowers and ferns. He had previously indulged his appreciation for floriculture while governor of New York, the position he held immediately before becoming president. The governor's mansion in Albany boasted a small but well-stocked conservatory where Cleveland began the lazy tours that he would continue in the White House's much larger facility. The president looked to this time in the glasshouse for both relaxation and amusement, and it was just about his only form of exercise. Weighing more than 250 pounds, he was not fond of sports, hiking, or even long walks.

After a yearlong courtship, President Cleveland wed Frances Folsom, the 21-year-old daughter of his late friend and former law partner Oscar Folsom. The intimate ceremony took place on June 2, 1886, in the Blue

First Lady Lucy Webb Hayes spent much time in the Conservatory, and was there with her children one winter day in 1879 when Theodore Davis, an artist for *Harper's Weekly* newspaper, arrived with his daughter, Carrie, and photographer G. W. Patch. The first lady agreed to sit for a photograph, and little Carrie (wearing a white apron) bashfully joined her and her children Fanny and Scott in front of the lens. On this particular day Mrs. Hayes was searching out the best ferns in the collection so that their likenesses could be reproduced on a specially designed dessert service. As she explained to Davis, the goal was to relate the dessert course to the "Conservatory Promenade" that immediately followed it. The artist liked the idea very much but suggested she go a step further: why not decorate an entire banquet set with drawings of American plants and animals? Mrs. Hayes wholeheartedly agreed and decided on the spot that Davis would be the illustrator. The State Service was introduced in 1880. Among Davis's vast array of designs were soup bowls adorned with corn and tomatoes, fruit plates with apples and blueberries, and dinner plates with deer, turkeys, and other wild animals. The Hayes china includes some of the most interesting service items on display in the White House collection.

Room. Henry Pfister had the place gleaming with flowers brought in from the Conservatory and greenhouses. Hundreds of potted plants, forced into bloom, overflowed tables and decorated the mantel. Scarlet begonias simulated flames inside the fireplace, while centaurea appeared as ashes on the hearth. Everywhere else stood an almost unbelievable assortment of roses, ferns, smilax, lilies, pansies, tulips, white narcissus, and orchids. The air was heavy with a rich floral perfume as the bride and groom exchanged vows. After the ceremony, they and their guests retired to the State Dining Room for dinner, where Pfister's skill was once again on display, as was his wit. The head table was decorated with a three-masted ship made entirely of flowers that the gardener had christened the *Hymen*.

By this time camellias had been a staple of White House flower decorations for more than fifty years, and the demand for their large, vibrant blooms was greater than ever. The old greenhouse, formerly Andrew Jackson's orangery, was now the camellia house and grew nothing else. The entire glasshouse complex, however, was in need of a serious overhaul. The yearly government reports echo the same refrain: despite regular repairs that included everything from painting to cleaning the boilers to replacing rotten wood and broken glass, they really needed to be replaced with new buildings erected around iron skeletons. As in the past, Congress declined, but did provide three greenhouses with new wooden frames and gave the orchid house a new roof, with a sloping, curvilinear pitch, so it came to resemble a barn.

First Lady Ida Saxton McKinley, a kindly but frail woman who suffered from epilepsy, mostly stayed out of the public eye and found contentment in the Conservatory. She was certainly on good terms with Pfister, who presented her with a rose bouquet each day. Although the stocks in the two rose houses had to be expanded to keep up, the garden staff efficiently refashioned the bouquets into new floral arrangements to decorate the mansion. For dinner parties, the table was heaped with roses, whether or not Mrs. McKinley herself was feeling well enough to attend. At a public reception just before the outbreak of the Spanish-American War in April 1898, Pfister

Frances Benjamin Johnston took this photograph of the White House Conservatory and sprawling greenhouses c. 1898. This image depicts the arrangement of buildings shown in the 1899 plan with the rebuilt orchid house visible at the extreme left.

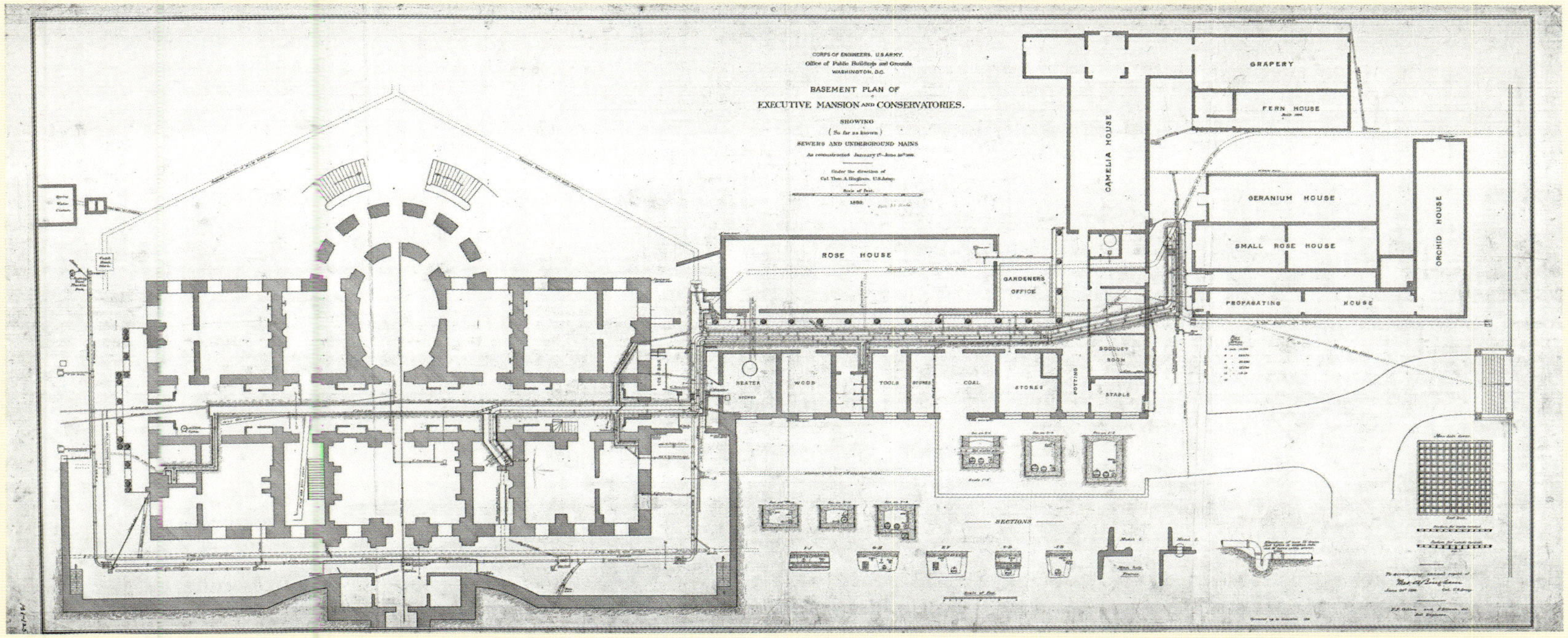
CORPS OF ENGINEERS. U.S.ARMY.
BASEMENT PLAN OF
EXECUTIVE MANSION AND CONSERVATORIES.
GRAPERY
FERN HOUSE
CAMELIA HOUSE
GERANIUM HOUSE
SMALL ROSE HOUSE
ORCHID HOUSE
ROSE HOUSE
GARDENERS OFFICE
PROPAGATING HOUSE
SECTIONS

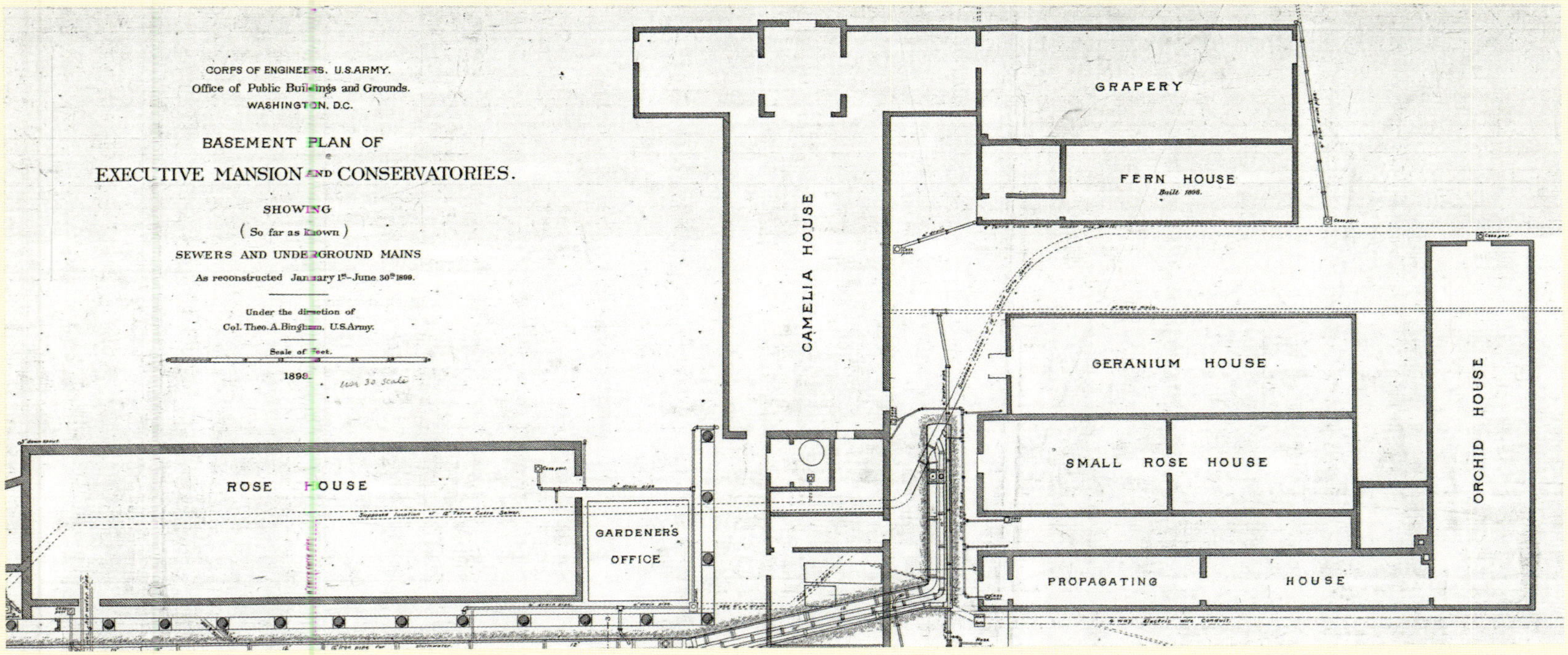
CORPS OF ENGINEERS. U.S.ARMY.
Office of Public Buildings and Grounds.
WASHINGTON. D.C.
BASEMENT PLAN OF
EXECUTIVE MANSION AND CONSERVATORIES.
SHOWING
(So far as Known)
SEWERS AND UNDERGROUND MAINS
As reconstructed January 1st–June 30th 1898.
Under the direction of
Col. Theo. A. Bingham, U.S.Army.
Scale of Feet.
1898.
GRAPERY
FERN HOUSE
Built 1896.
CAMELIA HOUSE
GERANIUM HOUSE
SMALL ROSE HOUSE
ORCHID HOUSE
ROSE HOUSE
GARDENER'S OFFICE
PROPAGATING HOUSE

GM
CENTRAL TRUCK
301-322-9000
ISUZU

In preparation for White House events, boxwood topiary are unloaded from a delivery truck (opposite) and a florist prepares centerpieces for a State Dinner in 2010.

By the spring of 1903, McKim was planning to erect a new, much smaller conservatory on the South Lawn of the White House as conciliatory gesture to the first lady, who was distressed over the loss of so many rare and valuable plants. Mrs. Roosevelt initially approved of the new conservatory, which she described as "a little pleasure palace," but when the cost of the White House renovation threatened to outstrip the congressional appropriation, it was one of the first luxuries that she dispensed with. By that fall, the proposal had been abandoned, and the final nod toward any kind of greenhouse on the White House Grounds was over.

Today, more than a century later, photographs of the conservatory and its group of greenhouses recall an era of hothouse gardening that has been replaced by outdoor gardens. Presidents now host events in the Rose Garden, and guests enjoy views of the Jacqueline Kennedy Garden, both nestled close to the mansion. White House greenhouses still provide fresh flowers for the President's House, as well as ornamental arrangements that serve as backdrops for speeches and special events. These buildings are all off-site, however, currently located at the far eastern corner of the District of Columbia near the National Park Service's Kenilworth Park and Aquatic Gardens. Trucks make regular deliveries, shuttling large quantities of plant materials to and from the White House. With a little advance notice, plants can be forced into bloom, thereby providing flowers on demand and out of season.

The White House itself also maintains a limited collection of live plants. Clipped trees planted in tubs soak in the sun shining through large windows, much as their predecessors did in Andrew Jackson's orangery. Laurel, in particular is an enduring favorite, and ficus trees are not uncommon. The "orchid craze" begun by Julia Grant never completely left the White House, although the numbers of individual plants generally changes from administration to administration. Finally, showcased in antique vases and fresh-cut bouquets, there are always roses.

EPILOGUE

"The History of the White House Grounds Is Hard to Trace"

THE FUTURE

NINE MONTHS AFTER THE CLOSE OF HIS PRESIDENCY, WILLIAM HOWARD TAFT succinctly stated that the "history of the White House grounds is hard to trace." He was very much correct; there had, as yet, been no serious study devoted to the subject. More than a century has since passed, and the landscape is now far better understood and far better appreciated than it was in Taft's day. This is, of course, precisely as it should be, and the preceding pages have endeavored to codify facts, clear up long-standing misconceptions, and present both the physical and social history of the President's Grounds. The history of the White House Grounds, however, probably lies as much in its future as in its past. Unlike many historic places, the White House is rooted in the past without being frozen in time. Its historic period is not limited to the eighteenth century, or the nineteenth, or the twentieth, or even the twenty-first. Instead, it is wonderfully open-ended, leaving plenty of time for future presidents, landscape professionals, and, yes, generations of everyday Americans to continue to shape, use, and appreciate the most famous yard in America. If the past is any indication, this compelling story will surely continue for many years to come.

grounds, because I would first wish to submit them for the approval of the President."

Downing drowned on July 28, 1852, before he had finalized his plans for the President's Grounds. He was most likely on his way to Washington when his steamboat's boiler overheated and exploded on the Hudson River. Although he had no meaningful influence on the President's Grounds, gardeners John Saul and John Watt implemented some of what he had planned elsewhere on the public grounds—primarily in the Ellipse and Lafayette Park, which although subsequently altered may be the most complete extant example of Downing's ideas. Saul and Watt remained in Washington for many years and themselves influenced the development of the President's Grounds in other ways. Saul, an Englishman whose brother worked as manager and part owner of Downing's old nursery business, went on to found a large nursery of his own just beyond the city's border and supplied trees, vegetables, and other plant stock to Watt, who served as head gardener at the White House from 1852 to 1862.

John Watt
Gardener (1852–62)

Scottish immigrant John Watt (1824–1892) was appointed White House gardener in February 1852, at the personal request of Andrew Jackson Downing, for whom he had worked for a short time. John Saul, Downing's foreman, had known Watt for years and evidently vouched for him. It is likely that Watt met Downing's need for someone who would not question his orders and would faithfully execute his plans for the President's Grounds.

Watt barely began work before Downing died, and little is known of his activities until the summer of 1853, when President Franklin Pierce authorized him to overhaul the President's Grounds. This work seems to have focused on maintaining and tidying up the landscape; it was not the thorough renovation Downing envisioned. A white picket fence was put up around the ornamental garden on the southeast, and its arbor was rebuilt and painted to match. The benches under this lattice were painted dark green and set upon freshly graveled walks. Watt also purchased vegetable seeds for the kitchen garden and ornamental plants for the White House conservatory and greenhouse. Much of this new stock came from his old associate John Saul.

An air of corruption has long hung over John Watt's name. Shortly after taking on the position, he was accused of embezzling government funds, but managed to convince his supervisors that he had made honest accounting errors brought on by inexperience in handling large sums of money. He was in trouble again in 1858, and the charges were serious enough to warrant several federal employees submitting sworn affidavits. The secretary of the interior, who had the final say on all matters relating to Washington's Public Grounds, eventually ruled that "Watt's defense is conclusive of his innocence."

Although the extent of the gardener's shady practices will likely never be known, it is a fact that he aided First Lady Mary Todd Lincoln in disguising her excess spending on clothes, food, furniture, and decorations for the White House. He accomplished the deceit primarily by moving money between accounts, all the while assuring the first lady that doing so was "right & legal" and actually a common practice. Watt also paid people for work that was never done. The White House's French cook, Auguste Julien, suddenly showed up on the rolls as a laborer on the grounds, as did a coachman and a butler. At one point Watt had a bill paid by selling a large rug right off the White House floor. It was promptly replaced at the public's expense. A State Dinner that cost $300 was billed out at $900. Perhaps most brazenly, Watt bought two cows and attempted to hide the purchase in an appropriation for manure used on the public grounds instead of appropriately charging the animals against the White House expenditures. The rationale was apparently that the cows would indeed produce manure.

More scandals followed. Watt was accused of being a Confederate sympathizer and spy. When Mary Lincoln publicly defended him, she herself fell under suspicion. After all, the first lady was from Kentucky and had five brothers serving in the Confederate Army. Questions were also raised regarding the nature of her relationship with Watt. He always seemed to be in her company, and wild rumors began to circulate that the two were having an affair. The final straw came in December 1861, when President Lincoln found his unreleased State of the Union message published in the *New York Herald* newspaper. As the only copy was then sitting on his desk awaiting its transmittal to Congress, it was clear that someone within the White House had leaked its contents. Watt was quickly identified, and this time even Mrs. Lincoln refused to come to his rescue. He was dismissed from the post of White House gardener in early 1862.

Henry Pfister
Gardener (1877–1902)

On May 1, 1877, President Rutherford Hayes appointed Henry Pfister (1846–1925) as head gardener at the White House. A 52-year-old Swiss locksmith and machinist from the president's hometown of Cincinnati, Pfister seems to have known little about gardening at the time, but he was a successful businessman and Hayes valued his managerial skills. The president apparently brought him in as a sort of horticultural director instead of a hands-on gardener. Pfister soon hired an entirely new staff, all of whom did possess practical gardening knowledge. By 1880, his workforce included nine men and one woman, who was employed as a full-time bouquet maker.

Pfister evidently learned on the job, and by the 1890s had become a horticulturist of considerable skill and renown. He was particularly interested in growing freshwater plants at the White House and became an early pioneer of aquaculture in the United States. He shared this passion with First Lady Frances Cleveland, and together they stocked the fountains with various species of flowering water lilies. Pfister also supervised the expansion of the White House glasshouses, which at their peak included nine separate buildings plus the Conservatory. Many of the plants gown in the greenhouses were set out for display in the Conservatory and used in cut-flower arrangements that decorated the mansion's interior. Each year thousands of annual plants were also grown in the greenhouses and later planted outside on the President's Grounds.

Pfister's tenure ended in November 1902. The greenhouses and Conservatory were gone, removed to make way for the West Wing, and the plants he had cared for, which were supposed to be transferred to new greenhouses, died from exposure. Broken-hearted and no doubt feeling betrayed, he departed quietly.

Frederick Law Olmsted Jr.
Landscape Architect (1893–1949)

The Olmsted name is synonymous with landscape architecture. Frederick Law Olmsted Sr. (1822–1903) gave birth to the profession, and his namesake son enhanced it. Washington, D.C., boasts many of Frederick Law Olmsted Jr.'s (1870–1957) most celebrated projects: the National Zoological Park, the National Mall, Rock Creek Parkway, Theodore Roosevelt Island, the Jefferson Memorial, and the grounds of the National Cathedral. In 1901, President Theodore Roosevelt appointed Olmsted a member of the Senate Park Improvement Commission for the District of Columbia, best known as the McMillan Commission after its chair, Senator James McMillan. In this capacity, Olmsted played a major role in devising a comprehensive plan that updated Pierre Charles L'Enfant's unrealized vision of the national capital for the twentieth century. The design of Washington today owes much to Olmsted's skills as a city planner, a vocation that he would shape and steer throughout the 1910s and 1920s.

Olmsted was also a conservationist, and on his 83rd birthday the Olmsted Grove in California's Redwood National Park was dedicated in recognition of his determination to protect the giant trees. His commissions for the National Park Service included work at Acadia, the Everglades, and Yosemite, where he served on the committee of experts from 1928 until his death in 1957. Olmsted wrote the mission statement for the 1916 National Park Service Organic Act, which brought the National Park Service into being, and in so doing laid out the agency's philosophical mission and legislative mandate: "To conserve the scenery and the natural and historic objects and the wild life therein and to provide for the enjoyment of the same in such manner and by such means as will leave them unimpaired for the enjoyment of future generations."

Finally, Olmsted was a historic preservationist, although the term itself would not become commonplace until after Congress passed the National Historic Preservation Act of 1966. In the interim, the visionary Olmsted crafted his now-famous plan for improving and maintaining the President's Grounds. On one hand it is a practical document designed to alleviate problems relating to automobile traffic, presidential privacy, security and safety, trees and other plantings, and views to and from the presidential mansion. On the other hand, Olmsted also carried out one of the earliest comprehensive preservation management programs in the United States. His core tenet, that the historic aspects of the President's Grounds should be preserved while simultaneously accommodating the current and future demands upon the landscape, may seem inherently obvious, but it was groundbreaking for its time. Even today such a delicate balancing act is the exception to the rule rather than the rule itself, and many other important historic sites have not been afforded the same protection.

Beatrix Jones Farrand
Landscape Gardener and Architect (1895–1955)

Beatrix Jones Ferrand (1872–1959) was briefly educated in Berlin in 1893 before studying horticulture and landscape design under Charles Sprague Sargent, the first director of Harvard University's Arnold Arboretum in Boston. Sargent encouraged Farrand to become a professional, and she never forgot his dictum "to make the ground fit the plan and not twist the ground to fit the plan." Following a European garden tour, she established her own landscape architecture office in New York in 1896. In 1899, she was the only woman in the group of ten professionals who established the American Society of Landscape Architects, and she remained an active member throughout her career. She considered herself a landscape "gardener," rather than "architect," a word she felt "should be left to the designers of buildings."

Eleanor McPeck, Farrand's primary biographer noted her "subtle and harmonious approach to color" and described her design style as combining "horticultural impressionism and the best elements of classical European gardening." Farrand was an artist who painted on landscapes instead of canvases. Indeed, taking a tour of one of her surviving gardens is about as close as anyone will ever get to stepping into a painting. Topography, vegetation, and materials play off one another to create special places that are distinct and imaginative, yet somehow also delicate and restrained. It is therefore no surprise that First Lady Ellen Wilson, an artist herself, selected Farrand to redesign the East Garden of the White House in 1913. Both believed that a garden of any size "must be treated in the impressionistic manner."

Although neither her East Garden nor her earliest gardens survive, one of Farrand's most successful designs, done for Abby Aldrich Rockefeller in Seal Harbor, Maine (1925–50), is still maintained by the Rockefeller family. Also extant is Dumbarton Oaks (1921–47) in the Georgetown neighborhood of Washington, D.C., which she transformed from a former farm into one of the most ingenious and artistic gardens in the United States. Farrand also created and implemented garden plans for the University of Chicago (1919–36) and Yale University (1922–45), as well as Hamilton (1924), Vassar (1926–7), and Oberlin (1939–46) colleges. She devoted the last years of her life to establishing the Reef Point Gardens on her family's property on Mount Desert Island, Bar Harbor, Maine.

Rachel Lambert Mellon
Landscape Designer
Rose Garden (1962–63) and Jacqueline Kennedy Garden (1964)

Privileged with great wealth, the late philanthropist Rachel Lambert Mellon (1910–2014) enjoyed the rare opportunity to devote her long life to her greatest passion: design. Bunny, as she was commonly known, "believed that style was order and that order brought pleasure," according to a recent biographer. Although she received no formal training, Mrs. Mellon designed the interiors for the dozen-plus residences that her family owned in the United States, France, and the Caribbean. They showcased a world-class art collection, but she preferred relatively small, intimate houses that were intentionally unostentatious, "livably elegant," and "clear of pretension." Professional designers took note of Mrs. Mellon's remarkable rooms, as did First Lady Jacqueline Kennedy, who enlisted her help in redecorating the White House interiors. According to longtime *Washington Post* journalist Maxine Cheshire, no one had a greater influence in shaping Mrs. Kennedy's celebrated sense of style and taste.

Bunny Mellon's friendship with the Kennedys led to her creation of the Rose Garden and Jacqueline Kennedy Garden outside the White House. These commissions were not, however, from a sense of obligation, and her work was a donation. Although technically an amateur, she had been an active student of gardening and horticulture from an early age. Her lifelong fascination began as a child while watching gardeners from the Olmsted Brothers company tend to the grounds of her family's home in Princeton, New Jersey. She also loved to read about gardens, both real and of the fairy-tale variety, and studied prints from old books. Her earliest forays into garden design took the form of miniature models that incorporated "small stone steps, real soil, and tiny topiary trees" fashioned from "sponges, glue, wires, and wood."

As an adult, Mellon's work was most strongly influenced by the seventeenth-century French gardeners, André Le Nôtre and Jean-Baptiste de La Quintinie, both of whom were employed by King Louis XIV to design the Royal Grounds surrounding the Palace of Versailles. She in turn created the landscapes for many of her family's properties, including the French-inspired gardens at her Oak Spring estate in Upperville, Virginia. At the age of 23, Mrs. Mellon undertook one of her first commissions, a garden for fashion entrepreneur Hattie Carnegie. She subsequently produced more landscape plans for private residences and public spaces, including the Kennedy family home in Martha's Vineyard, the John F. Kennedy Presidential Library in Boston, and River Farm, the headquarters of the American Horticultural Society in Alexandria, Virginia. In France, she helped fashion mogul Hubert de Givenchy design his gardens at the Château du Jonchet and assisted with the restoration of the Potager du Roi, the king's kitchen garden at Versailles.

Irvin Williams
Chief Horticulturist (1962–2008)

Irvin Williams first came to work at the White House during the Eisenhower administration. The president had recently been honored by having a new species of camellia named after him: the General Dwight Eisenhower Camellia. The flowers of this hearty shrub emerge a brilliant flame-red each spring, and in 1956 Williams planted an excellent specimen near the South Portico as well as a grouping of twelve matching plants near the western mound on the South Grounds. At the time he was not yet part of the official White House staff, but rather a National Park Service horticulturalist working out of an office at the Kenilworth Park and Aquatic Gardens in Northeast, Washington, D.C. Nonetheless, his great skill and expertise quickly made him an indispensable participant in the care and management of the President's Grounds. Before long, he knew more about the historic trees and other plantings than anyone else.

At the request of Rachel Lambert Mellon, Williams was transferred to the White House to assist with design and installation of the Kennedy Rose Garden, and he remained the authority on all of the plants growing on the President's Grounds for the next forty-seven years. He executed Mrs. Mellon's plans for the Rose Garden and Jacqueline Kennedy Garden in the 1960s. When she returned to update the gardens in 1981, Williams again provided his assistance, and ensured that all the flowers that First Lady Nancy Reagan had specifically requested were added to Rose Garden, among them her own Nancy Reagan Hybrid Tea Rose and the Patricia Nixon Floribunda Rose.

Many other projects were less memorable but still vitally important. A good example occurred in 1965, when it was discovered that the luxuriant Merion bluegrass on the South Lawn was suffering from Fusarium blight, a fungal disease that also harms wheat and oats. The diagnosis was hampered by the fact that the effects of Fusarium blight strongly resemble those of drought, and Washington, D.C., had experienced a long spell of dry weather. Upon detecting the true cause of the trouble Williams was able to treat the grass and bring it back to form.

Williams retired in 2008, by which time he had far and away become the longest-serving gardener in the history in the White House. The regular care and maintenance that he bestowed upon the landscape for so many years allowed it to become the perfect background for countless photographs and an ideal setting for official functions and private gatherings. Williams has been perhaps the site's greatest steward, and his seemingly endless knowledge and experience were surpassed only by his kindly disposition, good nature, and love of the President's Grounds.

Dale Haney
Gardener and Supervisory Horticulturalist (1972–2008)
Grounds Superintendent (2008-present)

Dale Haney has worked on the President's Grounds since 1972, beginning as a National Park Service gardener shortly after earning a degree in horticulture from Sandhills Community College in Pinehurst, North Carolina. Throughout his years of service Mr. Haney was promoted multiple times, and served as grounds maintenance supervisor and supervisory horticulturist. In addition to his many duties relating to the trees, flowers, and other plants growing on the grounds, he has also helped care for every presidential dog since King Timahoe, Richard Nixon's Irish setter. In 2008, Mr. Haney accepted a position with the Executive Office of the President as the Superintendent of the White House Grounds. Today he supervises a full-time staff of approximately twelve NPS gardeners and maintenance professionals who tend to the upkeep and maintenance of the presidential landscape.

254, 261
knee-wall, 30
knocking, 142, 178
knolls, 48
Kodak, 46
Korean spice viburnum, 90

L

L'Enfant, Pierre Charles, 2–3, 5–6, 12, 15, 118
  plan, 2–3, 5, 15, 118
laborers, 8, 12, 15, 51, 52
labyrinthine conglomerate, 276
laces, 137, 146
Lady Anne's Blush, 266
lady-slippers, 40, 261
lady watch, 156
Lafayette Park, 20, 22–23, 43, 53, 121–22, 176, 186, 188, 203, 207
lamps, 32, 37
  glass oil, 30
  large iron gas, 32
land, 2–3, 5, 12, 16, 20, 52, 121, 156, 199
  acquisition, 5
  invasion, 174
  ownership, 5
  vacant and unimproved, 121
landscape, 5–6, 11–12, 15, 18, 20, 29–30, 64, 174, 184, 186, 198, 203–4, 206–7, 261, 263
  historic, 115, 158
  ornamental, 206
  pock-marked, 256
  unbroken, 12
landscape architects, 18, 20, 88, 103
landscape plan, 12, 15–16, 51, 53–54, 71, 174, 194, 248
  ambitious, 15
  new, 18
landscaping, 48, 138, 258
Lane, Harriet, 268, 270
larkspurs, 78, 102
Latrobe, Benjamin Henry, 12–13, 15, 44, 48, 51, 144, 194, 214
  designs, 12–13
latticework pergola, 82
laurel, 194, 283
  mountain, 72, 198
Lauren, Lauren Bush, 75
lavender bushes, 102
lawn, 18, 57, 126, 129, 134, 142, 164, 181, 188, 262
  central, 90, 95, 106
  grassy, 113, 188
  green, 48
  open, 90, 194
  sloping, 23
  southern, 165
lawn mowers, 184
Le Blond, Ellie, 74
Le Blond, Sam, 74
Lean Bear, 270
Lebanon, 172, 237
Lee, Robert E., 179
Leinster House, 6
lemons, 264
level lawn, 48
Levy, Uriah P., 38
Lewis and Clark Expedition, 122
Lewis, Meriwether, 122
Liberty Bonds, 184
Library of Congress, 208
Lichtenstein, Roy, 108
light, 12, 48, 126, 148, 186, 198
  blue, 148
  electric, 92
  incandescent, 148
  largest, 148
  varicolored, 151
lilacs, 113, 194, 258
lilies, 42, 88, 258, 261, 275
lily pads, 75
lima beans, 252
lime, 200, 203, 264
limited trap-and-release program, 158
Lincoln, Abraham, 37, 64, 81, 97, 128, 130, 133, 140, 162, 164, 176, 178–79, 181–3, 204, 251–52, 270
  administration, 37, 64, 97, 133, 178, 251
  Civil War, 176, 178–79, 181–83
  emancipation, 178
  family, 162, 164, 176, 181–82, 183, 251
  park, 81
  tree, 204
Lincoln, Mary Todd, 164, 181–82, 251, 270
  conservatory, 270
Lincoln, Robert Todd, 162
Lincoln, Tad, 162, 164, 182
Lincoln, Willie, 176
linden tree, 208, 243
  little-leaf, 220
link, 3, 6, 12, 71, 226, 228
Linnaean Hill, 258
lion's foot, 250
liqueurs, 273–74
little-leaf linden, 240
live plants, 248, 250, 266, 270, 283
  setting tubs of, 42
lizards tail, 250
loads, 15, 115
  hundred wagon, 204
loan, 184, 270
location, 2, 20, 53, 72, 75, 156, 158, 206–8, 211, 226, 268, 280
  new, 266
  undetermined, 16
Logan, 81
Lombardy poplars, 8
Long Island, 203
lounging, 122
  roughly dressed laborers, 126
low-sloping glass-paned roofs, 266
Lowe, Truman, 108
Lowell, Robert, 152

M

Macaroni, 63, 168
macaw, blue, 166
MacKenzie, Peter, 266, 268
made-to-order, 266
Madison, Dolley, 174, 197
Madison, James 15, 52, 174, 194, 197–98, 248, 251
  administration, 197
  chief steward, 197
  Montpelier, 198
  slaves, 174
Madison Place, 52
Madonna lilies, 97
Magic Flute, 151
magnolias, 90, 198 231
  Jackson, 228, 231
  mature, 231
  saucer, 232
  stately southern, 228
  young southern, 231
Maher, James, 203, 250
Maher's nursery, 203
Maine, 103
maintained canal, 121
maintenance, 18
  costs, 270
  deferred, 38, 84
  seasonal, 20
Malayan sago, 264
Manassas, Va., 179
manpower, 8, 184, 204
Manship, Paul, 108
mansion, 6, 8, 11–12, 15–16, 18, 113–14, 146, 148, 174, 179, 181, 206, 261–62, 273, 275–76
manure, 203, 264
maple, 200, 203, 204, 220
maps, 2, 48, 75
  topographical, 50
marble fragment, 158
march, 134, 179
Marguerite, Anne, 197
Marine Band, 126, 129–30, 133–34, 137, 140, 144, 146, 148
Marine barracks, 172
marshy area, 52
marshy swamplands, 68
Maryland, 2, 156, 174
Massachusetts, 248, 256
Massachusetts Historical Society, 16
Massachusetts state senator, 198
master plan, 18, 198
masterwork Sylva, 199
McClellan, George, 183
McGrath, Thomas, 174, 197, 248
McKee, Benjamin Harrison (grandson), 46–47, 164–65
McKee, Benjamin Harrison (president), 46–47, 164–65
  grandchildren, 46–77, 164–65
McKee, Mary Lodge, 47
McKerichar, Alexander, 203
McKim, Charles Follen, 18, 113–14, 276, 280, 283
McKinley, Ida Saxton, 113, 275–76
McKinley, William, 113, 137, 275–76
  assassination, 276
McMahon, Bernard, 194, 248
Meade, George, 183
Meadville, 181
meals, healthy 254
medicinal properties, 199
medieval apothecary, 250
Meigs, Montgomery C., 180
Mellon, Rachel Lambert, 88, 90, 98, 106, 108, 234
  selection of plants, 88
memoirs, 133, 266
memorials, 71, 118
memorial spaces, 3
memory, 97, 200, 226, 250
Mendelssohn, Felix, 148
menus, 153, 274
Meridian, 81
metals, reinforced, 30
Michler, Nathaniel, 121
Mid-Century East Garden, 104
Mid-Century West Garden, 84
middle class, 126
Mikado, 134
military, 133, 186
  camaraderie, 150
  engineer, 2
  exercises, 174, 179
  families, 143
  hospital, 182
  officers, 126
  parade, 125
  purposes, 172
  strategists, 114
militia, 8, 178
milk, 60, 166, 246
Millet, Francis, 151
Mills, Robert, 261
miniature farm, 251
minister, 137, 198, 261
Miriam's Kitchen, 254
miscellaneous garden plan, 16
Mitchell, D. W., 126
mixed-use garden southeast, 258
model, 174
  combat, 184
Mondo grass, 72
monkey flower, 250
Monroe, James, 30, 50–51, 197–98, 264, 266
  administration, 50–51, 198
  mounds, 50–51
Monticello, 13, 15, 38, 194, 248
Montpelier, 198
monumental architecture, 3
monumental entrance arch, 174, 197
Mooley Wooly, 246
morality tales, 148
morning glories, 270
mounds, 48, 51, 57, 186, 188, 203
  earthen, 51, 57, 140, 188
  east, 51
  eastern, 48, 51, 208
  encaustic tiles past, 274
  Jefferson, 48, 50–51
  Monroe, 51
  western, 48
mounted cannon, 174
Mount Pleasant Triangle, 81
Mount Vernon, 18, 78, 224, 264, 270
Mozart, Wolfgang Amadeus, 151
muddy quagmire, 36
mudslinging, 228
mulberry, 194, 200
Mullett, Alfred B., 52–53, 60, 62, 113–14
Munger, George, 174
music, 133, 146, 148, 152
  classical, 153
  military, 133
  popular, 133
  sheet, 134
musical accompaniment, 137, 146
musical apprenticeship, 134
musicales, 133, 153
musicians, 133, 151
myth, 228

N

Nacotchtankes (people), 2
Nanko, 164
Nanny, 164
narcissus, 72, 81, 258, 261, 275
nation, 8, 11–12, 20, 38, 108, 128, 137, 142, 152–53, 165, 172, 250, 261, 273
national anthem, unofficial, 137
National Archives, 48, 51, 208, 251, 262
national capital parks, 20, 104
National Christmas Tree, 186, 188
National Gallery, 108
national gardening trade, 203
National Intelligencer, 270
National Mall, 71, 121, 198
national monuments, 118
National Park Service, 20, 75, 88, 104, 156, 188
  plan, 53
  Kenilworth Park, 283
Native Americans, 2, 95, 122, 270
  chiefs and tribal elders, 122
naturalistic pattern, 103
naturalizing, 250
natural life, 199, 214
natural resource stewardship, 97
natural woodland, 207
navy, 8, 38
Navy Building, 52
Navy Department 12, 261
navy searchlight, 148
neat wrought-iron picket fence, 60
negotiations, 5
neoclassical purity, 276
Neshanock potatoes, 250
net, 156
Netherlands, 104, 198
Netherlands Carillon, 104
Nevelson, Louise, 108
newspapers, 122, 128, 146, 154, 158, 165, 214, 228, 273, 275
  accounts, 165
  announcements, 133
  articles, 204
  editorials, 46
  seller, 188
New Year's Day, 11, 182
New York, 22, 30, 121, 125, 146, 165, 258, 274, 280
New York City, 262
New York Evening Post, 154
New York's Old City Hall, 3
Nicolay, John, 270
Nicotiana, 90
Nixon, Richard, 97, 108, 152–53
  administration, 108
Noah's ark, 125
Noguchi, Isamu, 108
nomenclator, 200
Norcross Brothers, 280
North Drive, 22, 36–37, 217, 243
northeast, 37
  level ground, 248
Northeast Gate, 180, 220
Northeast Washington, 88
North Elevation, 6
North Entrance, 11, 30, 36, 38
  drive, 40, 114
  entranceway, 12
North Facade, 12, 18, 36
North Fence, 179
North Fountain, 40, 42–43, 64, 67
North Front, 6, 15, 36, 38, 179, 197, 206
North Gate, 174
North Grounds, 5, 15, 18, 29–30, 32, 36, 38, 40, 42, 47, 125, 129, 194, 204, 206
  accessible, 46
North Lawn, 6, 122, 125, 162, 174, 179, 197, 204, 206, 208
North Portico, 22, 36, 38, 40, 180, 183, 206–7
Northwest Gate, 22, 33
Norway spruce trees, 203
nozzle, 67
Nugent, Lyndon Patrick, 72, 74
nurseries, domestic, 273
nursery, 199–200
  dedicated, 156
  small, 261
nurserymen, 250
  finest, 194
nursery receipts, 208
nurses, 47, 126, 140, 165
nuts, 222

O

oaks, 18, 199–200, 203, 224
  chestnut, 200, 214
  cork, 200
  largest, 148
  pin, 222
  scarlet, 220
  species of, 199, 222
  white, 200
  willow, 200, 236, 238
  willow-leaved, 194
Obama, Barack, 97, 143, 156, 160, 162, 168, 170, 220, 243, 246, 254
  administration, 254
  family, 168, 170
  gardens, 254
Obama, Malia, 97, 168, 170
Obama, Michelle, 109, 153, 170, 188, 220, 254
Obama, Sasha, 97, 168, 170
occasions, 97, 129, 162, 165, 168, 179, 214, 250, 266
  historic, 11
office, 12, 15, 20, 60, 72, 84, 92, 125–26, 137, 151, 162, 165, 170, 264, 266
  gardener's, 276
  housed, 280
  left, 15, 248, 261
  open air, 92
  outdoor, 92
  presidential, 81
  president's, 26, 76
officer, 162
  career, 38
  commanding, 178, 182
  disbursing, 164
  federal, 121
  volunteer, 178
office space, requisite, 114
office wing, new, 276
Ogle, Charles, 250
Ogle's speech, 261
Ohio buckeyes, 220
  planted, 22
O'Keeffe, Georgia, 108
Oklahoma City, 172, 242
O'Laughlin, John, 165
old-fashioned carnival, 152
old-fashioned flowers, 78, 258
Old World, 12
Olmsted Jr., Frederick Law, 18–20, 23, 47, 57–58, 60, 71, 84, 203–4, 206–7, 226
  plan, 18–19
  revisions, 60
  vision, 207
One-Half Street, 52
open spaces, 88, 115, 118
  green, 42
operatic adaptations, 133
Operation Squirrel, 158
orangery, 264, 266
  new, 264
  relocated, 263, 270
orchestral scores, 134
orchestral suites, 152
orchid collection, 273
  new, 273
orchid craze, 283
orchid culture, 273
orchid fever, 273
orchid house, 273, 275–76
orchids, 273, 275
orchid varieties, new, 273
orders, 126, 179, 181–82, 203, 248, 251, 270, 273
  issued, 181
  official, 138
  requisite, 15
Oriental spruce trees, large, 186
ornamental garden, 57, 60, 250, 256, 261–62, 266, 268
Osmanthus ilicifolius, 90
Oswego County, 125
Ousley, John, 200, 208, 250–51, 258 261, 264, 266
outbuildings, permanent, 12
Outdoor Oval Office, 92
outdoors, 151, 153, 168
  cherished time, 168
  concealed positions, 178
  executive branch, 92
  hotbeds, 264
  space, 84
outhouse, 11
Oval Office, 92, 168, 170, 232, 236, 276
Oval Room, 11
overcrowding, 206
overhang, 114
  sheltered, 36
overlapping iron arches inset, 33
overlook, 203, 250
Oyster Bay, 146, 165
oyster supper, 182

P

painting, 11, 68, 152, 174, 176, 197, 200, 231, 266, 275
pair, 12, 15, 36, 48, 57, 164–65, 181, 186, 188, 203, 228
palace, 2, 6, 12, 15
  a little pleasure, 283
  aborted, 12
  king's, 6
palings, 32–33, 35
palms, 78, 264, 266, 268, 274
  sago, 270
Pan, 82–83, 86
Pandora, 148
panic, full-scale, 179
pansies, 81, 261, 270, 275
parade ground, 179, 182
  planned, 71
parades, 2, 179
parcels, 2–3, 5, 125, 248
pardoning, 97
  presidential turkey, 97
Paris, 38
park, accessible, 115
parklets, 118
parks, 118
  city's, 203
  public, 20, 118, 121–22, 126
  urban, 3
park systems, 118
  modern, 121
parsnips, 248, 250
  yellow-flowered, 250
parterres, 40, 90
  diamond, 90
  formal, 113
  geometric, 100
  pinwheel-shaped, 40
Pasadena, 103
passages, 12, 250
  short, 273
  slow, 20
passageway, open, 273
patch, 231, 248, 275
  abused, 15
  ragged, 121
  small, 250
Patent Office and Botanic Garden, 268
path, 203, 266
  dirt, 12
  flagstone, 72
  horseshoe-shaped, 274
  original, 121
patriotic songs, 133
patterned gardens, 78
Patterson, Martha, 246
payroll records, 51
peace, 90, 276
peach, 194, 200
peas, 248
  marrowfat, 250
Pebble Beach, 22
Pei, I. M., 106, 109
Pen-lu sod, 156
Pennsylvania, 72, 181–83, 222, 250
Pennsylvania Avenue, 3, 6, 14–15, 22–23, 29–30, 32, 43, 60, 181, 248, 261
  crossed, 15
  marched down, 179
Pennsylvania Railroad, 8
Pennsylvania Volunteer Infantry Regiment, 179
people's house, 26, 46
performances, 130, 134, 137, 146, 148, 151
  best theatrical, 148
  outdoor, 151
performers, 148, 152
perimeter, 47, 60, 62, 86, 113, 115
  garden's, 102
  mansion's, 6
  wall, 12
periwinkle, 72, 88, 90, 258
persimmons, 200, 264
Peters Jr., Richard, 222
  Belmont, 222
Peters, Roberta, 152
pets, 162, 165–66, 168, 246
petunias, 40, 90, 97, 270
Pfister, Henry, 42, 67, 203–4, 208, 210–13, 273, 275, 280
  greenhouses, 273, 275
  trees, 208, 210–13
Philadelphia, Penn. 2, 6, 38, 222
Philadelphia Enquirer, 142
phlox, 40, 78, 90, 270
photographers, 47, 150, 274–75
  Mathew Brady, 270
  see also by name
photographic record, 231
photographs, 35, 38, 42, 46–48, 62, 67, 71, 78, 83–84, 128–29, 134, 138, 273, 275–76, 278
  hand-colored, 103
  historic, 113, 203
  official, 95
  period, 122, 204
  springtime, 109
Philip, Prince, 152, 240
pickle beets, 250
picnic, 129, 178
picnickers, 121, 125
picnic-style, 153
picture, 20, 126, 146, 270, 274
picturesque, 15, 204
piecemeal, 18
  down, 30
Pierce, Edward, 5
Pierce, Franklin, 51
  mounds, 51
Pierce, Jane Appleton, 266
Pierce, Joshua, 266
Pierre, Jean R., 178
Piers, 30
Pike Jr., Zebulon, 122
piles, 11
  large, 174
pine, white, 194
pink blossoms, 266
Pink Cloud, 90
pink organdy dress, 146
pinks, 102, 261
  china, 266
  mullein, 78
pipes, 11
  tin, 266
pit, 11, 160
  horseshoe, 160
  trash, 158
pitch, 160
  curvilinear, 275
  horseshoe, 160
plan, 2, 5, 12, 15, 18–20, 52, 54, 57, 67, 194, 203–4, 206, 208, 278, 280
  initial, 114

unrealized, 264
planned layout, 262
plantations, 178
planters, 22, 115, 256
tobacco, 5
planting beds, 103
arrow-shaped, 100
featured lateral, 106
straight, 81
planting list, 88, 194, 211–12
planting plan, 102, 113, 194
planting programs, 226
plantings, 20, 40, 42, 72, 76, 84, 113, 197, 199, 203, 206–7, 220, 222, 228, 240
beautiful seasonal, 20
close, 194
decorative, 261
mass, 207
new, 197–98, 206
ornamental, 256
single, 194
supplemental annual, 42
thick, 90, 197
plant materials, 283
plants, 88, 90, 103–4, 106, 108, 197–200, 248, 250, 258, 266, 268, 270, 273–74, 280, 283
aquatic, 42, 67, 270
archangel, 258
decorative, 251
delicate, 264
dusty miller, 40
edible, 248
fragile, 273
grafted, 200
hardy perennial flowering, 248
help, 254
hot-house, 248
large, 266
mental, 261
mock, 40
native, 199, 261
new, 250, 261
old-fashioned flowering, 100
ornamental, 76, 184, 262–63, 270
petunia, 261
procure, 250
rare, 274, 280
small, 264
specimen, 194
tubbed, 266
useful, 250
plateau, flat, 226
platform, 133, 140, 179
constructed, 178
elevated, 134
self-supporting, 168
wooden, 133
players, 142, 156
avid tennis, 154
international, 156
star, 158
playgrounds, 118, 121, 168, 170
children's, 138, 168
personal, 162
small, 170
pleasure garden, 54
plot southeast, 250
Plumbe Jr., John, 231
pocket change, 142
political adversaries, 46
political attacks, 228
political differences, 38
Polk, James K., 38, 130, 261
Pomroy, Rebecca, 164
ponies, 162, 164, 166, 168
calico, 168
pony rides, 152
pony shed, 168
pools, 64, 102, 144
large rockwork water, 273
rectangular lily, 103
swimming, 20, 23, 156, 158, 160
poplars, 206, 214, 226
tulip, 214, 226
poppies, 78, 258
monster, 148
porte cochere, 114
expanded, 114
portrait, 14, 174
Portugal, 198
post-and-chain fence, 36
post-and-rail fence, 12, 199
simple wooden, 12
wooden, 200
post oak, 200
posts, 32, 35, 162, 180, 203, 280
potatoes, 166
Potomac River, 2, 12, 52, 64, 67, 68, 104, 176, 181
potted plants, 42, 264, 266, 275
Prato, Rodica, 22
precision marching, 126
presidential aide, 274
presidential election, 228
presidential enclave, 18
presidential landscape, 6, 18, 48, 52, 71, 122, 129, 207, 264
Presidential Palace, 2
President's Grounds, 11–13, 15–16, 18, 20, 128–30, 164–65, 172, 174, 199, 203–4, 206–8, 214–15, 246, 261–62, 279–81
President's House, 3, 5–6, 8, 11–12, 14–15, 30, 46, 48, 174, 197, 244, 262, 283
see also White House
President's Levee, 125
President's Palace, 3, 5, 15, 20, 250
President's Park, 20, 52–53
president's physician, 156
press, 148, 165–66, 228, 231
press area, 22
Preston, William, 197
pride of helen, 102
primroses, 90, 258, 261
prince, 250, 270
Prince Albert peas, 252
Princeton University, 151
privacy, 46–47, 144, 158, 204, 207, 266, 268, 280
outdoor, 44, 47, 129, 146
presidential, 26, 29
problem
restored, 47
screens, 81
securing effective, 47
private courses, 156
private donations, 158
private gatherings, 137
private-public foundation, 153
private quarters, 29
private setting, 158, 268
privet, 47, 88, 194
privy, 8, 11
prized possessions, 264
proclamation, 174
procured star jasmines, 266
projects, 3, 5, 12, 52, 54, 64, 81, 88, 103, 106, 114, 176
earthmoving, 5, 51
monumental construction, 84
promenaders, 121, 137, 258
gentlemen, 273
propagating house, 276
pruning-knife, 274
Prussia, 198
Public Grounds, 18, 122, 128–29, 138
puppies, 168, 182
Puritan minister, 168
Purple Luster, 90

Q

Qimmiqsak, 108
Queen Elizabeth II, 152, 240
Queen Juliana, 104
Quincy, Mass., 256

R

races, 125, 129, 138, 143
impromptu, 164
racial equality, 178
racial segregation, 140
radio broadcasts, 158
radish, 248, 251
rag-a-muffins, 125
railing, 30, 36
rain, 26, 128, 138
swift splashing, 40
ranking officers, 181
ranks, 48, 162
fictitious, 162
raspberries, 250, 252
ratification, 2
Reagan, Ashley Marie, 74
Reagan, Cameron, 74
Reagan, Nancy, 106
Reagan, Ronald, 106, 115, 142–43, 231
rebuilding effort, 174
receptions, 104, 130
documented, 8
first-ever outdoor, 151
formal, 122
large, 266
personal, 178
presidential, 114
public, 151, 275
recognition, 106
recommendations, 18, 20, 206–7
reconfiguration, 114
reconstruction, 51, 113, 174, 256
recreation, 121, 144, 154, 182
Red Cross, 184
red twigged lindens, 203
Redmond, Robert, 86, 226
reefer jackets, 140
Reef Point Gardens, 103
reelection, 11, 200, 228, 250
Reeves, Bill, 280
refugee beans, 252
relaxation, 92, 274
relic hunters, 204
relief, 251, 266
sharp, 148
religious gathering, 178
relocate, 2, 261
remainder, 47, 133
reminds, 199, 254
remodeling, 22, 113
renaissance, 194
renderings, 12
renovated maintenance building, 168
renovation, 84, 104, 106, 266, 276, 280
rental properties, 11
reorganization, 20
replacements, 86
reporters, 150, 165
reputation, 154, 266
natural, 3
resemblance, 78, 168
reshaping, 52
residence, 11, 22–23, 178
official, 26
permanent, 199
private, 262
Residence Act, 2, 8
rhododendrons, 113
Rickey, George, 108
River Farm, 32–33
roads, 54, 57–58, 60
associated, 57
east-west running, 57
national, 106
new, 52
system of, 54, 57
transverse, 54, 57
roadways, 52, 54, 160
50-foot-wide, 52
distinctive guitar-shaped, 57
envisioned, 54
new, 52
new east-west, 54
robber barons, 140
Rock Creek Park, 143
Rollo, 166
Roman motifs, 3
roof, 71, 270, 273
new, 174, 270, 275
tile, 264
rooms, 64, 114–15, 126, 133, 156, 261, 266, 270, 275
bouquet, 276
central, 268
dining, 29
dressing, 29, 148
fumigation, 270
interior, 8, 76
living, 29
Robb, Lucinda Desha, 72, 74
Roosevelt, Alice, 146, 154, 165
Roosevelt, Archie, 165–66
Roosevelt, Edith, 78, 90, 98, 100, 146, 168, 280, 283
garden, 78
renovation, 280, 283
Roosevelt, Eleanor, 188
Roosevelt, Ethel, 165, 168
Roosevelt, Franklin Delano, 18, 20, 47, 57, 114, 150–151, 170, 186, 203, 226, 262, 266
trees, 226
war, 186
Roosevelt, Kermit, 165–66
Roosevelt, Quentin, 165–66, 168
Roosevelt, Theodore, 18, 42–43, 47, 137, 142, 146, 148, 154, 156, 165–66, 168, 224, 276, 280
Beaux-Arts, 42
family, 146, 165–66, 280
renovation, 18, 276, 280, 283
Sagamore Hill, 146, 280
tennis court, 156
trees, 224
Roosevelt Jr., Theodore, 165
rooster, 166, 168
roots, 203, 250
bulbous flower, 248, 258
esculent, 250
hundred asparagus, 250
rope ladders, 170
roses, 78, 84, 90, 261, 264, 268, 270, 275, 283
bushes, 84, 90, 261
magnificent, 113
namesake, 108
new floribunda, 86
pink, 146, 276
red, 276
smaller hybrid tea, 84
white, 276
white tree, 97
rosette, 35
route, 12, 115
horseshoe-shaped, 273
row, 106, 199, 206, 222, 231, 266, 268, 270, 274
banked, 266
long, 88
stately, 18
straight-line, 204
Russia, 198, 224
Russian tree, 224
Russo-American Oak, 224

S

sacrifices, 184
safekeeping, 174
safety, 26, 158, 179
Sagamore Hill, 146, 280
Saint Bernard, 166
sale, 184, 248, 251–52
Salem Witch Trials, 222
Salvation Army, 134
salvia rings, red, 22
Sand Creek Massacre, 270
sandstone shell, 12
santolina incana, 90
saplings, 174, 200, 206
Saturday concerts, 133–34, 137
Saul, John, 251–52, 266, 270
Scala, Francis, 133
scalloped urn, 35
scarlet begonias, 275
scarlet locust, 194
scene, 8, 32, 46, 71, 125–26, 148, 183, 188, 197, 204, 248
desolate, 8
idealized, 122
jubilant, 178
picturesque, 15
serene, 146
schoolchildren, 254
local, 148, 254
Scotch, 194
screen, 47, 194, 207, 258
latticework, 262
sculpture, 106, 108, 152
Seale, William, 174
searchlight, 148
sea serpents, 64
season, 133, 137, 199, 266, 283
holiday, 186
seasonal flowers, 106, 115
seasonal replanting, 42
seats, 2, 148, 266
seclusion, 47, 268
Second Floor, 133
secretary, 8, 165, 188, 198, 224, 250
displeased, 60
Secret Service, 186
security, 26, 30, 47, 60, 128, 137, 153, 156, 172, 179, 186
growing, 115
tightened, 172, 186
security barrier, 62
security measures, 142
Sedium sieboldii, 90
seedlings, 199, 222, 250
seeds, 200, 222, 224, 248, 250–51, 270
sycamore, 222
trading, 261
vegetable, 248, 251
warehouse, 251
senators, 92, 126, 176, 197–98
sentinels, 207
sentries, 162
service buildings, 12, 81
276
ancillary, 244
services, efficient, 179
servicewomen, 143
settees, cast-iron, 261
sewer, 121
shade, 18, 47, 92, 178, 204, 224, 273
shagbark hickory, 200
Shakespeare, William, 148
Shapiro, Joel, 108
Shasta daisies, 90
shellbark hickory, 200
shelter, 264, 280
shelves, 188, 266, 268
flying offstore, 168
shepherd's purse, 250
Sherman, William T., 183
shoes, 140
canvas, 137
kid, 140
shrubs, 15, 18, 90, 118, 194, 197, 203–4, 211–12, 231, 248, 258, 261–62
flowering, 104, 266
large, 208
sidewalk, 29, 37, 114
sightline, 6, 18, 68, 71, 203, 207
cardinal, 71
primary, 206
signatures, 51, 182
Silsbee, Nathaniel, 222
silver urns, 174
Sioussat, Jean-Pierre, 174
sixteen-gun salute, 8
Sixteenth Street, 15
skeleton, inner, 6
skilled tradesmen, 8
slaughterhouse, 121
slaves, 8, 129, 176, 178
slide, 168, 170
slope, 52, 54, 138
natural north-south, 12
Smith, David, 108
Smith, John, 2
Smith, John Rubens, 44
Smith, Andrew C., 270
Smithsonian Castle, 121
smokehouses, 12, 244
snapdragons, 78
snow, 170, 231
social elite, 126, 266, 273
society, 32, 103, 106, 126, 146
soil, 5, 104, 200 *see also* dirt
soldiers, 150, 151, 153, 162, 174, 178–79, 181–84, 188
Sousa, John Philip, 134
South Carolina, 197
South Drive, 46
southeast, 57, 60, 194, 258
garden area, 264
massive chestnut oak, 208
southeast corner, 15, 44, 174, 197
southeast entrance, 14, 46
southern addition, 52
southern boundary, 13, 54, 57
southern magnolia, 228, 231
Southern Plains delegation, 270
South Facade, 62
South Fountain, 67, 148, 186, 226
South Front, 18, 104, 231, 261
South Grounds, 15–16, 46–48, 53–54, 57, 62–64, 67–68, 71–72, 126, 128–30, 133–34, 144, 146, 148, 156, 203–4
expanded, 60
extended, 53
lower, 67
new, 52
private, 125
South Lawn, 47–48, 128–29, 138, 140, 146, 148, 151–54, 156, 166, 176, 182–84, 186, 188, 215, 217
South Portico, 67–68, 76, 78, 81, 98, 125, 128, 130, 140, 142, 146, 148, 151, 231, 262
South Portico steps, 97
South Vista, 68, 71–72, 203, 207
southwest, 44, 57, 60, 261
gully, 174
space, 8, 12, 16, 44, 72, 76, 90, 95, 108, 113, 115, 126, 129, 138, 142
aesthetic, 273
ceremonial, 95
decorative, 262, 273
divided, 12
enclosed, 98
expanded, 266
functional, 15, 29, 266
living, 280
narrow, 270
private, 95, 270
public, 29, 106, 122, 203
sheltered, 47
verdant, 121
spades, 220
spandrels, left, 33
Spanish-American War, 275
Spanish broom, 194
Spanish chestnut, 200, 222
spatial division, 250
species, 42, 88, 197, 200, 203, 206, 208, 211, 214, 231
fast-growing, 174
favorite, 261
slow-growing, 214
tropical, 273
specimen rosebushes, 81
specimens, 203, 208, 228, 240, 250, 266, 270
historic, 204
killed delicate, 266
prized, 280
smaller, 220
speeches, 95, 133, 151, 152, 250, 283
spikes, 35
spirea, delicate, 270
sports, 118, 144, 154, 156, 274
spotlight, 46, 148
Spring Garden Tour, 137
spruce pine, 194
spruces, 188, 203, 231
square wooden boxes, 266
squirrels, 158, 166
stabilization, 84
stables, 12, 244
staff, 104, 154, 176
domestic, 162
new gardening, 203
president's, 276, 280
Stafford County, Va., 6
stage veteran, 148
stairs, 15, 113
Standing in the Water, 270
stars, 3, 40, 90, 148, 266, 276
state, 11–12, 26, 52, 172, 174, 184, 198, 220
native, 220
official, 104
wild, 11
State Arrival Ceremony, 137
State Department, 273
State Dining Room, 174, 264, 273
State Dinners, 114, 130, 152, 254, 266, 283
State Floor, 26, 128, 273
State Service, 275
statue, 38, 40, 43, 71, 82
Stayman winesap apple tree, 72
Stephenson, Isaac, 246
Stephenson's farm, 246
steps, 18, 36, 47–48, 88, 92, 114–15, 148, 151, 166, 264, 273, 275
borderline mutinous, 182
stone, 270, 273
stereograph, 29, 37, 40, 64, 146
Sterro-Photo Company, 276
Stewart, John, 40, 42
Stewart's grand planting scheme, 42
stocks, 203, 250, 266, 275
growing, 251
Stoddert, Benjamin, 8
Stone, Charles P., 178–79
Stone Jr., Edward Durrell, 72
Stone, James, 30
stone, coarse, 12
stone coping, 204
stone gateposts, 22, 204
Stonehenge, 22
stone piers, 15, 30
stone wall, 15, 44, 54, 68, 174
blackened exterior, 197
storehouses, 6, 12, 244
storm, 130, 134, 203–4, 208, 240
St. Petersburg, 224
strawberries, 250, 251, 264
strolling, 122, 126, 274
strolls, 26, 274
structural repair, 84
structure, 5–6, 113, 174, 266, 273, 280
ancillary, 12
finished, 268
grand neoclassical, 261
massive, 113
monumental, 15
narrow, 273
new, 266
rectangular, 114
temporary, 8
two-story, 114
structure brick, 266
Stuart, David, 5
Stuart, Gilbert, 174
Sumner, Charles, 224
stylized creatures, 64
Suffruticosa, 90
sugar maples, 194, 203
summer, 11, 22, 32, 40, 42, 88, 92, 151–53, 174, 176, 248, 262, 266
downpour, 148
sun, 133, 181
hot summer, 179

southern, 264, 273
sunburst pattern, 15
sunk, 57
flat-roofed hyphens, 12
grassy expanse, 18
sunken ditch, 68
superintendent, 88, 208, 236
longtime, 192
on-site construction, 8
suppliers, 203, 258, 266
supplies, 30, 104, 222, 246, 264, 270
supporters, 125, 133, 178
Suwanee River, 142
swamps, 2, 5, 224
sward, 133
sweet alyssum, 266
Sweet William, 102
swift waters, 68
swing, 156, 168, 170, 182
symmetrical flower parterres, 78
symmetrical landscape treatment, 29
symmetry, 15, 113
system, 266, 270
compost, 254
expansive, 57
online lottery, 143
timed ticketed, 172

T

table
bedside, 198
matching, 72
overflowed, 275
presidential, 250, 270
president's, 246
stepped, 274
taffeta, white, 146
Taft, Helen, 81, 148, 151, 246
Taft, William Howard, 47, 137, 148, 151, 156, 246
tank, 184
official Liberty Bond, 184
tapestry, beautiful, 98
tapis vert, 12, 16
tarragon, 250
tastes, 15, 42, 108, 133, 152
Taylor, Zachary, 126, 130
temperance movement, 273
tennis, 154, 156
court, 23, 156, 160
regulation, 154
tournament, 156
tennis ball lettuce, 254
Tennis Cabinet, 154
tents, 148, 152, 181
flag-draped army, 146
fly, 92
striped, 152
temporary, 122
terraces, 12–13, 72, 75, 110, 125, 270
flagstone, 72
terrace wings, 244
terrorist attacks, 97, 115, 172
Thanksgiving turkey, 97
theater, green, 88
thick groves, 18, 203, 204
thinning, 71, 206–7
third-generation globe trekker, 224
Thomas, John T., 166
Thornton, William, 8, 248
thoroughfares, 3, 52, 54
three-masted ship, 275
Tiber Creek, 52, 68, 121
tickets, 143, 148
Tidal Basin, 71, 81
tobacco plantation, 5
tomatoes, 188, 275
topiary hollies set, 106
topographical survey, 12
topography, 48, 206
topsoil, 104, 110
tornado, 203–4
touch-me-not, 250
tour, 115, 126, 172, 270, 274
tradition, 42, 122, 130, 137–38
annual, 142–43
lighthearted, 97
long-standing, 220, 258, 266
traditional dances, performed, 122
traffic, 115
trampoline, 168
Treasury, 12, 52, 183, 261
Treasury Building, 22–23, 52, 81, 113, 148, 261, 266
completed, 261
construction superintendent, 261
expanding, 266
grounds, 20
hulking, 110
new, 261
tree branches, 103
tree canopy, 226
tree culture, 198
tree house, 168, 170
tree-lined allées, 15, 81
tree nursery, 199–200, 222, 250, 258
tree-planting programs, 198, 200, 206
tree plantings, 71, 194, 203–4, 206, 220, 226
tree population dates, 208
trees, 18, 148, 164–65, 192, 194, 197, 199–200, 203–4, 206–8, 211–12, 214, 220, 222, 224, 231–32
apple, 222
best, 203
collected, 222
commemorative, 220, 228, 232
cork, 215
deciduous, 203, 231
descendant, 224
disfigured, 204
eastern, 186, 188
ficus, 283
flowering, 194
foreign, 250
forest, 198–99, 250, 258
gallows, 222
grown, 113
historic, 222
immature, 206
large, 182, 194, 203
lemon, 270
long-lived, 206
lost, 220
magnificent, 20
mangled, 204
mature, 54, 158, 207, 236, 240
new, 20, 200, 204, 206–7
oak, 198
orange, 270
ornamental, 197, 231, 262
preexisting, 203
relocated, 203
shade, 22, 198
smaller, 194
strong, 178
surviving, 220
tulip, 194, 200
western, 188
young, 200, 206, 222, 224
Trenton Evening Times, 148
trespassers, 60
tribal elders, 122
troops, 142, 178–79, 270
tropical paradise, 273
tropical plants, 264, 273
Truman, Harry S., 84, 88, 104, 106, 143, 226
renovation, 84, 86, 104, 143, 226
trumpet creeper, 261
trunk diameters, 214
trunks, 148, 203, 207, 214
tubs, 125, 268
tulips, 104, 248, 258, 261, 270, 275
turkeys, 97, 275
Tuscan columns support, 12
Twelfth Street, 261
Tyler, John, 125–26, 130

U

underground cisterns, 266
underground supply pipe, 67
understory, 206–7
uniforms, 148
Union, 128, 133, 176, 178, 183, 220
Union cattle ground and slaughterhouse, 121
Union loyalists, 178
Union troops, 68, 162, 164, 178–179, 182, 183
United States, 36, 38, 158, 160, 174, 176, 184, 194, 197, 224, 226, 248, 250, 264, 266
United States Army Old Guard Fife and Drum Corps, 137
United States Botanic Garden, 198

V

Van Buren, Martin, 250, 261
vault, 264
fireproof, 266
fireproof Treasury, 264
masonry, 68
massive groin, 15
vegetable gardens, 188, 246, 248, 258
vegetables, 15, 248, 250, 254
verdant greensward, 11
Veteran Reserve Corps, 182
Victoria lettuce, 252
Victorian, 18, 76
Victory Garden, 188, 254
Vietnam, 152–53
vines, 16, 208, 261, 268, 270
Virdis Marginata, 90
Virginia, 2, 6, 32, 104, 179, 226, 256
Virginia cedars, 103
virgin's bower, 250
vision, 5–6, 15, 20, 81, 137, 174
visite, 64, 68
visitor gateway, primary, 110
visitors, 5, 11, 15, 26, 29, 98, 108, 113, 118, 122, 126, 128, 133, 172, 178
captured, 46
vista, 18, 207
vote, 182–83

W

Waddell, Peter, 6, 68
Wagner, Richard, 134
wagon, 164–65, 174
Wales, 270
walkways, 15, 258, 273
curving, 113, 258
special, 81, 83
wall, 15, 30, 52, 60, 174, 273
exterior, 84
heavy masonry, 264
high, 12
high rock, 15
new, 12
north, 276
red brick, 12
stone plinth, 35
west, 273
walled ditch, 12
wallflower, 258
Walter, Thomas U., 261
walnuts, 200, 222
war, 52, 92, 97, 151–53, 172, 174, 176, 181–83, 186, 188, 197, 203, 254, 256, 261
War Bonnet, 270
wartime, 116, 151, 172, 183, 184, 186
Washington, 2–3, 5–6, 8, 38, 81, 121–22, 125–26, 146, 178–79, 181–83, 197–98, 203, 214, 250–51, 266
departed, 179
Washington Circle, 81
Washington Evening Star, 137, 140
Washington, George, 2, 5–6, 8, 18, 38, 57, 71, 78, 81, 125, 142–43, 148, 174, 203, 207, 222, 224, 264, 266, 268, 270, 280
Mount Vernon, 18, 57, 78, 224, 264
orangery, 270
tomb, 224
trees, 224, 266, 268
Washington guidebook, 48, 51
Washington hawthorn, 90
Washington Metropolitan Police, 179
Washington Monument, 18, 71, 121, 148, 203, 207, 280
Washington Monument Grounds, 142–43
Washington Playgrounds Association, 148
Washington's command, 2
Washington's military hospitals, 251
Washington's orangery, 270
Washington's parks, 121
Washington's tomb, 224
watch, 46, 140, 179, 181
water, 40, 64, 103, 270
pressurized, 40
pump, 270
running, 266
water jets and sprays, 40, 64, 67, 148
water lilies, 42, 67
watermelon, 252
water pressure, 40
water skyward, 113
water springs, 40
water tanks, 270
water trough, 184
watercolor, 8, 30, 46, 48, 81, 90, 174
Watt, John, 251, 266, 268, 270
Watteau type, 146
Waud, Alfred, R., 176, 179
weather, 130, 143, 148, 266, 268
bad, 114
cold, 12, 264
dry, 36
foul, 36
warm, 72, 75, 126
Wedding March, 148
weeds, 248
weeping willows, 194, 197
Welsh Terrier, 168
Westchester County, 22, 30
West Colonnade, 22–23
western expedition, 122
western section, 276
western wing, 273
West Executive Avenue, 22–23, 52, 60, 204, 251
west facade, 110
West Garden, 24, 76, 78, 80–82, 84, 88, 90, 92, 95, 103
West Terrace, 12, 15, 148, 261, 264, 266, 268, 270, 273, 276
colonnaded, 81, 95
surviving, 113
West Wing, 22–23, 53, 81, 83, 92, 154, 158, 168, 246, 261, 280
expansive, 52
Wheeler, Perry 88
Whelan, William 250
whiskey punch, 125
White, Douglass, 266
White House, 11–13, 15–16, 18–19, 29–30, 42–44, 46–48, 113–15, 125–26, 150–54, 164–66, 178–83, 197–200, 203–4, 261–64, 273–76
White House Grounds, 47–48, 50–52, 53, 54, 71–72, 121–23, 125–27, 143, 151–55, 157–59, 163–68, 172–75, 178–79, 181, 183–89, 207–8
White House Liaison, 20
White House receptions, 109, 268
White House Sculpture Garden, 108
White House Visitor Center, 109, 222
white oak sapling, 224
William Pierce's Linnaean Hill Nursery, 258
Williams, Irvin, 88, 192, 208, 236
Williams, Moreley Jeffers, 18
willow oak trees, new, 115
Willis, Nathaniel Parker, 125
Wilson, Edith, 103, 151
Wilson, Ellen Axson, 76, 80–81, 83–84, 90, 98, 103, 151
gardens, 80–1, 83–4, 103
Wilson, Woodrow, 81, 92, 95, 103, 137, 151, 164, 184
administration, 137
outdoor oval office, 92
windows, 125, 133–34, 148, 181, 264, 273, 283
bedroom, 81, 133
lunette, 12
narrow, 273
Windsor Castle, 261
Winningstadt cabbage, 252
winter, 42, 181, 204, 215, 264, 266, 268, 270
bitter, 264
frost, 266
greenery, 88
nights, cold, 264
quarters, 181
scene, 231
Wisconsin, 246
Wister, Langhorne, 182
wooded park area, 72
World's Fair, 90
World War I, 92, 137, 184
World War II, 84, 114, 143, 151, 186, 254
Wood, Robert, 32, 38
wrought-iron brackets, 30
Wyeth, N. C., 8

Y

Yale, Eli, 166
Yellow Wolf, 270
young dahlias, 258